I0045476

Anonymous

Historical and Technical Papers on Road Building

Anonymous

Historical and Technical Papers on Road Building

ISBN/EAN: 9783744648752

Printed in Europe, USA, Canada, Australia, Japan

Cover: Foto ©berggeist007 / pixelio.de

More available books at **www.hansebooks.com**

BULLETIN No. 17.

U. S. DEPARTMENT OF AGRICULTURE.

OFFICE OF ROAD INQUIRY.

HISTORICAL AND TECHNICAL PAPERS

ON

ROAD BUILDING IN THE UNITED STATES.

COMPILED UNDER DIRECTION OF

ROY STONE,

SPECIAL AGENT AND ENGINEER.

WASHINGTON:
GOVERNMENT PRINTING OFFICE.
1895.

LETTER OF TRANSMITTAL.

U. S. DEPARTMENT OF AGRICULTURE,
OFFICE OF ROAD INQUIRY,
Washington, D. C., April 3, 1895.

SIR: I have the honor to transmit herewith a number of papers on road building in the United States, which contain important information of a kind that is much in request, and I recommend the publication of the same as Bulletin No. 17 of this office.

Very respectfully, yours,

ROY STONE,
Special Agent and Engineer.

Hon. J. STERLING MORTON,
Secretary of Agriculture.

3

CONTENTS.

ILLUSTRATIONS.

HISTORICAL SKETCH OF NATIONAL ROAD BUILDING IN THE UNITED STATES.

By RICHMOND STONE.

The act of Congress of April 30, 1802, which provided for the admission of Ohio as a State made three propositions, to become binding on the United States as soon as the Ohio convention should accept them, with the accompanying condition. Of these propositions the third, which alone is considered here, read as follows:

> That one-twentieth part of the net proceeds of the lands lying within the said State sold by Congress from and after the 30th day of June next, after deducting all expenses incident to the same, shall be applied to the laying out and making public roads, leading from the navigable waters emptying into the Atlantic to the Ohio, to the said State and through the same; such roads to be laid out under the authority of Congress, with the consent of the several States through which the road shall pass.

The annexed condition was that the convention should, "by an ordinance irrevocable without the consent of the United States," exempt all lands thus sold from taxation for a term of five years after that sale.

These provisions had been reported to the House by a select committee in the same form in which they were finally passed, except that the proportion of the proceeds of land sales applicable to roads was put at one-tenth instead of one-twentieth. In this report the committee had followed the recommendation of Mr. Gallatin, then Secretary of the Treasury. In a letter dated February 13, 1802, he tells them that "the roads will be as beneficial to the parts of the Atlantic States through which they are to pass, and nearly as much to a considerable portion of the Union, as to the Northwestern Territory itself. But a due attention to the particular geographical situation of that Territory and of the adjacent districts of the Atlantic States will not fail to impress you strongly with the importance of that provision in a political point of view, so far as it will contribute toward cementing the bonds of the Union between those parts of the United States whose local interests have been considered as most dissimilar."

After some opposition the article was adopted. In the Senate the proportion was cut down from a tenth to a twentieth. This and the other Senate amendments were accepted by the House, and the measure became law.

This act was followed by that of March 3, 1803, directing the Secretary of the Treasury to pay over to the State of Ohio 3 per cent of the proceeds of the lands of the United States lying within that State sold after June 30, 1802, to be applied in making roads within the State. The condition attached to the grant, in the act of 1802, had been accepted by the Ohio convention in an ordinance of November 29, 1802.

In the next Congress the matter was again brought up. There was some question whether the Ohio convention had really accepted the proposition of the act of 1802. It had asked that 3 per cent of the proceeds of the lands sold should be expended within the State and under the direction of the legislature. The point in dispute was whether the 3 per cent asked was additional to the 5 per cent offered by the act, or whether it was only a part thereof, of which the convention, in duly accepting the proposition, had requested control. It was finally amended, leaving only the undisputed 3 per cent to be expended on roads from eastern rivers to the Ohio, and in that shape passed. A bill was accordingly passed by which the President was authorized to name commissioners to explore eligible routes. In the Senate the bill was amended and, on its return to the House, laid over until the next session. At that session, however, it seems to have been crowded aside by the impeachment proceedings against Judge Chase.

In the Ninth Congress it was more fortunate. In the Senate Mr. Tracy, of Connecticut, brought in a report on the act of 1802, stating that the sales of Ohio lands since June 30, 1802, had amounted to $632,604.27, and recommending that a road be laid out from Cumberland, on the north bank of the Potomac River, to a point on the Ohio opposite Steubenville, and a little below Wheeling. A bill in accordance with this report was introduced and passed by the Senate. In the House it met with opposition; some members advocating three roads instead of one, others wishing to postpone action. The $50,000 appropriated by the Senate was cut down to $30,000, and with this and some other changes the bill finally passed. The House amendments were agreed to in the Senate, and on March 29, 1806, the bill was approved by President Jefferson.

It provided that the President should appoint a commission of three to lay out a road on the line above mentioned, 4 rods in width. When this was done they were to submit a report and plan, pointing out the parts most in need of attention, and estimating the probable expense of improving them. The President was further authorized to accept or reject the report (in whole or in part) to obtain the consent of the States through which the route passed, and, that being obtained, to proceed in applying the sum appropriated in building the road. The latter, in addition to the requirement of 4 rods width, was to have a raised carriage way in the middle of "stone, earth, gravel, or sand," ditches by the side, and no grade elevation greater than an angle of 5 degrees with the horizon. The other details were left to the President.

During the six years from 1810 to 1816 six appropriations were made for continuing the Cumberland road, amounting in all to $680,000, to be replaced out of the Ohio land fund already spoken of.

In 1817 an attempt was made to create a new fund for internal improvement. A bill was introduced in the House by John C. Calhoun to set aside for roads and canals the bonus and dividends received by the United States from its newly chartered national bank. In supporting it Mr. Calhoun said:

The manner in which facility and cheapness of intercourse add to the wealth of a nation has been so often and ably discussed by writers on political economy that I presume the House to be perfectly acquainted with the subject. It is sufficient to observe that every branch of national industry—agricultural, manufacturing, and commercial—is stimulated and rendered by it more productive. The result is to diffuse universal opulence. It gives to the interior the advantages possessed by the parts most eligibly situated for trade. It makes the country price, whether in the sale of the raw product or in the purchase of the articles for consumption, approximate to that of commercial towns. In fact, if we look into the nature of wealth, we find that nothing can be more favorable to its growth than good roads and canals.

Let it not be said that internal improvements may be wholly left to the enterprise of the States and of individuals. I know that much may justly be expected to be done by them; but in a country so new and so extensive as ours, there is room enough for all, the General and State Governments and individuals to exert their resources. Many of the improvements contemplated are on too great a scale for the resources of States or of individuals, and many of such a nature that the rival jealousy of the State if left alone might prevent. They require the resources and general superintendence of the Government to effect and complete them.

But there are higher and more powerful considerations why Congress should take charge of this subject. If we were only to consider the pecuniary advantages of a good system of roads and canals, it might, indeed, admit of some doubt whether they ought not to be left wholly to individual exertions; but when we come to consider how intimately the strength and political prosperity of the Republic are connected with this subject, we find the most urgent reasons why we should apply our resources to them. In many respects no country of equal population and wealth possesses equal materials of power with ours. In one respect we are materially weak. We occupy a surface prodigiously great in proportion to our numbers. The common strength is brought to bear with great difficulty on the point that may be menaced by an enemy. Good roads and canals, judiciously laid out, are the proper remedy. Let us, then, bind the Republic together with a perfect system of roads and canals.

The first great object is to perfect the communication from Maine to Louisiana. This may be fairly considered as the principal artery of the whole system. The next is the connection of the lakes with the Hudson River. The next object of chief importance is to connect all the great commercial points on the Atlantic with the Western States, and, finally, to perfect the intercourse between the West and New Orleans. There are others, no doubt, of great importance, which will receive the aid of the Government. The fund proposed to be set apart in this bill is about $650,000 a year, which is doubtless too small to effect such great objects of itself, but it will be a good beginning. Every portion of the community, the farmer, the mechanic, and the merchant, will feel its good effects; and, what is of greatest importance, the strength of the community will be greatly augmented, and its political prosperity rendered more secure.

Henry Clay also spoke in favor of the constitutionality and merits of the proposed act. As an instance of the value of improved roads, he stated that when the Cumberland road and the State road from Baltimore to Cumberland were completed, the journey from Baltimore to Wheeling would be reduced from eight days to three. The House, however, amended the bill so that it placed the actual carrying out of the work in the hands of the States, the Government retaining the general superintendence. In this form it was passed.

On March 13, 1817, President Monroe returned the bill to the House with a veto message, in which he said that though aware of the great importance of roads and canals, and of the signal advantage with which the National Legislature might exercise such a power as proposed, he did not believe it was constitutional, even with consent of the States. An attempt to override this veto failed of the necessary two-thirds majority.

After this Congress returned to its former method. In 1811 5 per cent of the net proceeds of sales of public lands in Louisiana had, as in the case of Ohio, been given to the State for levee and road building; in 1816 the same percentage of a similar fund had been given to Indiana for roads and canals; and in 1817, the like for the State of Mississippi. This was now followed in 1818 by the gift of 2 per cent of a similar fund to Illinois for roads leading to that State; in 1819, 5 per cent to Alabama; the same in 1820 to Missouri, and in 1845 to Iowa.

Meanwhile annual appropriations for the Cumberland road of sums to be replaced from the funds thus set aside in the States through which it passed were continued. In one year, 1819, over half a million was given, and when the last appropriation was made ($150,000 by the act of May 25, 1838) the sum total amounted to about $7,000,000.

In 1822 the regular appropriation for the Cumberland road was vetoed by President Monroe. Accompanying his veto message was a paper in which he entered exhaustively into the whole question of the power of Congress in the matter. The bill vetoed went further than the mere application of the money; it provided for the collection of tolls, and for the protection of the road from malicious injuries. This, said Mr. Monroe, was exercising attributes of sovereignty which the Federal Government did not possess. Examining the precedents formed in the road construction already undertaken by Congress, he concluded that in no case had it attempted to exercise eminent domain in securing right of way, nor raised revenue from the roads, nor protected them by penal laws. These things were not within the power of Congress. It had only appropriated money for road building in the Territories, and in States which had given their consent. This power, he believed, it did possess. As to the advantages of a system of internal improvement, he was "under a deep conviction that they were almost incalculable, and that there was a general concurrence of opinion among our fellow citizens to that effect." The powers and jurisdiction necessary for carrying out such objects could not even be granted by the States, nor received by the Federal Government, save by a constitutional amendment, and this he recommended.

The Cumberland road has long since passed out of the hands of the National Government. An act of 1834, in appropriating $300,000 for repairs and improvements on those portions of the road which ran through Maryland, Pennsylvania, and Virginia, provided further that when thus completed they should be turned over to the States in which they lay, and that the United States should not thereafter be subject to any expense in repairing them. By an act of 1848, the Indiana section was surrendered to the State; the same was done in Ohio in 1833, and in Illinois by an act of May 9, 1856.

CONSTRUCTION OF THE CUMBERLAND ROAD.

The first contract for the road was concluded in the spring of 1811, and covered only the first 10 miles. The contractors agreed to clear

FIG. 1.—Cumberland road near Chestnut Ridge Mountains.

the route of trees to a width of 66 feet, remove the stumps, level a roadbed 30 feet in width (increased later to 32), and to reduce the grades to a maximum of 5 degrees, in accordance with the gradations fixed by the commissioners who laid out the road. The sides of embankments were not to exceed a slope of 30 degrees. Sufficient ditches were to be provided. The roadbed was to be covered for a width of 20 feet with stone 18 inches deep in the middle and 12 at the sides; the upper stratum of 6 inches to be broken so as not to exceed 3 inches in diameter: the lower stratum not to exceed 7 inches. The contract price for the whole 10 miles amounted to $60,328.25, which, together with additional allowances, the salary of the superintendent, mason work, and bridging, made an average of $7,500 a mile.

FIG. 2.—Cumberland road. Steepest grade between Wheeling and Cumberland.

The specifications seemed to have been followed in the later work also, except that on the line west of the Monongahela the maximum grade was reduced to 4½ degrees. By subsequent contracts the road was surfaced with 2 inches of sand and gravel, and rolled. In the western portions of the road, where no stone was obtainable, it was built entirely of gravel. The average final cost of construction of the whole road east of Wheeling was $13,000 a mile.

The road has been thus described by one who wrote of it in 1879:

It was excellently macadamized; the rivers and creeks were spanned by stone bridges; the distances were indexed by iron mileposts, and the tollhouses supplied with strong iron gates. Its projector and chief supporter was Henry Clay, whose services in its behalf are commemorated by a monument near Wheeling. There were sometimes twenty gaily painted four-horse coaches each way daily. The cattle and sheep were never out of sight. The canvas-covered wagons were drawn by six or twelve horses. Within a mile of the road the country was a wilderness, but on the highway the traffic was as dense as in the main street of a large town. Ten miles an hour is said to have been the usual speed for coaches, but between Hagerstown and Frederick they were claimed to have made 26 miles in two hours. These coaches finally ceased running in 1853. There were also through freight wagons from Baltimore to Wheeling which carried 10 tons. They were drawn by twelve horses, and their rear wheels were 10 feet high.

From Cumberland to Baltimore the road, or a large part of it, was built by certain banks of Maryland which were rechartered in 1816 on condition that they should complete the work. So far from being a burden to them it proved to be a most lucrative property for many years, yielding as much as 20 per cent, and it is only of late years that it has yielded no more than 2 or 3 per cent. The part built by the Federal Government was transferred to Maryland some time ago, and the tolls became a political perquisite, but within the past year it has been acquired by the counties of Allegany and Garrett, which have made it free.

Fig. 3.—Cumberland road. The "S" bridge over the Big Buffalo Creek, Pennsylvania.

The Cumberland road is by no means the only one for which Congress has appropriated money. At the very start in 1806 the opening up of the Ohio route was balanced by a provision for a road from the frontier of Georgia leading toward New Orleans, and one from Nashville to Natchez. From that time until 1838, the total sum of over $1,600,000 was given for various roads in a succession of annual appropriations. Of this, $200,000 was expended in Florida; $286,000 was expended on roads leading from Detroit to Chicago and other points, and $206,000 on a road from Memphis " to William Strong's house on the St. Francis River in Arkansas." In 1845 there were a few small appropriations, and then from 1854 down to the civil war another period of activity, during which over $1,600,000 was laid out chiefly on roads in the Territories. After the war, in 1865, there were appropriations amounting to $140,000, but since then little has been done beyond roads and streets in the District of Columbia, and roads leading to the national cemeteries from neighboring towns. For the construction and repair of these cemetery roads, annual appropriations of $10,000 to $15,000 are still made.

In addition to the sums of money above mentioned, grants of land have from time to time been made to the States in aid of road building, and the labor of the United States troops has been employed. The troops when so engaged in the early days were given an extra daily allowance of 15 cents and a gill of whisky.

By an act of 1824 the President was authorized to appoint a board, to consist of two civil engineers and such officers of the Engineer Corps as he might detail. This board was to survey and make estimates on routes for roads and canals of national importance. A sum of $30,000 was given for this purpose. That part of the board's work which related to roads consisted principally in an examination of routes from Washington to New Orleans, Buffalo, and Cumberland; from Baltimore to Philadelphia, and from Detroit to Chicago. It also laid out a continuation of the Cumberland road as far as Zanesville.

In 1830 a bill was introduced in the House for a road from Buffalo to New Orleans by way of Washington. It was debated at great length, but finally failed.

At the same session was brought forward the Maysville and Lexington turnpike bill, authorizing a Government subscription to the stock of a turnpike company in Kentucky. It was attacked as being a local and not a national matter, but passed the House and also the Senate. On May 27 President Jackson returned it to the House with a veto message, in which he said that, at the utmost, Congress had power to appropriate money for national objects only, and that the measure before him was local in its purposes. Though personally friendly to internal improvements, he believed it to be inexpedient for the nation to act in that direction until the national debt was paid off. Even then he had doubts on the constitutional question, and preferred a distribution among the States in proportion to their representation in Congress.

A few days later he vetoed a similar bill for Government subscription to the Washington and Rockville Turnpike Company in Maryland and the District of Columbia.

There are two reports which are of great interest in connection with this subject. The first was submitted in compliance with a resolution of the Senate on April 4, 1808, by Albert Gallatin, then Secretary of the Treasury. In it the "early and efficient aid of the Federal Government" is strongly recommended for a comprehensive system of roads and canals. As to the former, he advocates, first, a turnpike from Maine to Georgia, 1,600 miles in length, costing on an average

$3,000 per mile; secondly, roads to Detroit, St. Louis, and New Orleans. These latter he thinks would be of more importance to the Government than to individuals, and need be cleared only enough to allow the passage of the mails, and, possibly, of stages, at a cost of about $200 per mile, or $200,000 in all. This, with the cost of the Maine-Georgia turnpike, would make a total of $5,000,000.

He divides the roads to be built into three classes:

First. Earth roads with a convex bed, drains, and ditches, costing, with bridges, $500 to $1,000 per mile. These he recommends for the poorer districts, for districts where materials are lacking, for the low country south of 36° latitude, where frost would not affect them, and for the country north of 41° latitude, where the roads are covered with snow in winter and transportation is by sleighs.

Second. Roads with a gravel surface 6 to 9 inches thick, costing $3,000 per mile.

Third. Roads 22 feet in breadth, with a top stratum 12 to 18 inches deep of hard broken stone, or broken stone and gravel, costing $7,000. This class he intends for districts where there is a heavy commercial traffic.

He further advises the adoption of broad tires, and that all grades be reduced to a maximum of 5 degrees, and, where practicable, to 3½ degrees.

The other report alluded to is one submitted to the House of Representatives January 7, 1817, by John O. Calhoun, as Secretary of War. It consists principally of investigations into the military value of roads and canals, and advocates their construction chiefly on that account.

In this connection may also be noted a letter from Governor Lewis Cass to the Secretary of War, dated November 29, 1817. He strongly recommends the building of roads in the Northwest, as from Wheeling to Sandusky, and thence on to Detroit. He asserts that it would greatly add to the defensive strength of the United States in that region, and that it would promote speedy settlement and the sale of public lands even so far as to be in the end pecuniarily profitable. He thinks "the time can not be remote when the policy of connecting the different parts of this vast Republic by great permanent roads will be felt and acknowledged; when such a policy shall banish local jealousies and discordant interests, shall furnish new and increased facilities for private industries, and shall add strength and wealth to the resources of the nation."

ROAD BUILDING IN OHIO.

By Hon. MARTIN DODGE,
President Ohio State Highway Commission.

The State of Ohio stands in the first rank as to her road improvements, both in number and quality. Beginning at the era of national road building, Ohio had a larger mileage of the famous Cumberland road than any other State, the road extending through her territory a distance of about 200 miles, from Wheeling, on her eastern border, through Cambridge, Columbus, and Springfield to the west. This was well built by the General Government, at a cost of about $7,500 per mile, and is in a fair state of preservation for the most part. But with the advent of the railroad national road building ceased, and all additional road building in the State since that time has been done by counties and municipalities.

The most extensive road improvements outside of the paving of streets in cities and villages have been made in the form of free turnpikes under the 1-mile and 2-mile assessment plan, by which all land outside of municipal corporations and within 1 mile or 2 miles is taxed according to the benefits. Whether the 1-mile limit or the 2-mile limit is taken depends upon the petition which accompanies the application. There is also a supplementary provision in the general law by which the entire property of the county may be taxed at a rate not to exceed 4 mills on the dollar to supplement the fund raised for free turnpikes under the 2-mile assessment law as referred to above. Before this can be done the proposition must be submitted to the voters of the county and favored by a majority. Under the operation of these laws a number of counties in the State have built many miles of turnpike of an excellent character, especially the counties of Shelby, Union, Hardin, Logan, Clark, Fayette, Preble, Franklin, Warren, Miami, Darke, and Hamilton, which have an aggregate of 6,000 miles of turnpike. So that, with the national road, the free turnpikes in various counties, and her extensive system of canals, the State has always been in the first rank as to the means of cheap transportation.

During the last two decades activity in the line of turnpike building has been gradually upon the wane, partly for the reason that the railroads have been able to give a cheaper rate of transportation than could be had over the free turnpike, and partly for the reason that land values and the profits of agricultural industry have been gradually diminishing during that period. But within the past two or three years there has been a sort of revival of the good roads question, and, it being felt that the farmers were too poor to bear the burden of improving the roads by assessing the entire cost upon the benefited areas adjacent to the improved roads, considerable special legislation has been enacted by the general assembly with a view to reviving the activity in road building. The special law enacted for Cuyahoga County is pernaps the most important of any. This county, though second in population and wealth, has never availed itself of the general provisions of the law for building free turnpikes, and the only improved roads it had before the passage of this special law were toll roads built of plank. Toll roads and toll bridges are generally of short duration among the American people. The time has come when their abolition is demanded and when free roads must take their place. To provide for this in Cuyahoga County, and also to build free roads where no toll roads had been built, the special law was passed in 1892. This act provides that all of the property in the county, both personal and real, shall be taxed at the rate of one-half mill on each dollar valuation on its assessed value; and in addition thereto the agricultural line outside of the city of Cleveland shall be taxed 1 mill on every dollar valuation. This is to form a general road fund for the improvement of the country roads in such manner as the county commissioners may direct.

No part of this fund is to be expended for bridges. The amount of the levy is about $80,000 per year, $60,000 of which is paid by the city of Cleveland alone. With this ample fund, which is more than is expended by the State of New Jersey under her new law, or of any other State for the direct improvement of roads, the commissioners of Cuyahoga County were better constituted to distinguish themselves by the improvement of our roads than any set of men have been since the General Government abandoned the construction of national roads. This magnificent sum of $80,000 is not an appropriation, but an annual levy continuing from year to year until the legislature shall repeal the law. Fully authorized, and well provided as the county commissioners were, they selected three separate roads leading in different directions from the city of Cleveland for improvement, and employed Mr. Jay F. Brown, civil engineer, to design and superintend the construction of these roads. They have already completed several miles, partly of brick and partly macadam. These roads are probably the best ever built in the State of Ohio, and Cuyahoga being a county peculiarly destitute of such road-building materials as are ordinarily used, it will be of great interest and benefit to other places similarly situated to know how these roads were built and what kind of material was used. I therefore submit a statement of plans, costs, and details, which Mr. Jay F. Brown. the engineer in charge, has kindly furnished me:

THE WOOSTER PIKE.

A description of the building of the road known as the Wooster pike will serve to illustrate how a good road may be made, over which heavy loads may pass at all times of the year, requiring but very little repairs for a long term of years, and those repairs being easily and cheaply accomplished.

The soil along that road is a heavy white clay, hard to drain and difficult to keep in place unless it is thoroughly graded and prepared to resist the action of frost or travel. It was claimed by many people who had spent their lives in the neighborhood that it would be impossible to so drain a road in that kind of soil that the water would disappear and the mud holes not occur. I think now, a year and a half since the road was finished, judging from the heavy travel in all kinds of weather, that the road is a complete success, and is a practical demonstration that a clay road can be drained so as to keep a uniform surface in wet weather. The drainage of that road was done in the following way:

The road originally was 60 feet wide from fence to fence. We graded the central part of the road, a roadway 32 feet wide. On each side of the roadway was made a storm ditch of an average depth of 4 feet, 2 feet wide on the bottom, slopes, and banks, the slope being 1½ feet horizontal to 1 foot vertical. After the roadbed had been brought to a grade line and thoroughly finished, a line of drainpipe, 6-inch capacity, was laid along each side of the 32 feet; that is to say, a trench was dug 16 feet from the center line of the street to a depth of about 4½ feet below the grade line of the roadbed. The trench, after laying the pipe, was filled with stone, broken

Fig. 4.—Showing method of holding brick in place alongside of dirt road.

to 2½-inch size. The pipe used was second quality of vitrified pipe, which can be procured very cheaply at the pipe factory. On account of 6-inch pipe being a standard size, more of it being used than of any other, there is always a large amount of what the engineers call "seconds" in the yard. The company is always willing to sell them very cheaply, and they answer for drainage purposes equally as well as first-class pipe. In fact, they are cheaper than the soft yellow draintile, which are liable to break and stop the flow of the water in the pipe, causing much trouble and expense for repairs. A drain of this kind was laid on each side of the roadbed, with outlets for water into every cross stream or ditch where it was possible to discharge the water, so frequent as not to overload the pipe in heavy storms. After the drains were put in, the strip of brick pavement above mentioned was laid close to one of the drains, leaving 21 feet width of dirt road for summer use. This dirt road was repeatedly rolled with a heavy roller until the upper foot or 2 feet of the crust of the roadbed became hard and solid. Our work on that road has demonstrated that heavy rolling of a road which has been properly drained will form a crust or a roof, so that the water can not stay on the road, but must flow at once into the drainpipe and disappear; and in case of storm water too rapid and too deep for the pipe trench to catch and carry off, the water flowed over the pipe trench into the storm ditch at the outside, which never fails to carry all the water that comes. Since the road was finished there has been no break, no settlement, no stoppage of water, no ruts, no mud, and travel on the road has doubled many times, thus showing the popularity of a hard, even roadway for winter travel as well as summer.

I think the building of the Wooster pike will settle the question as to the manner of getting a good road in a clayey country, where the transportation of material is from distant points, and consequently expensive; that is to say, the cost of hauling a ton of the best road material that can be had, like vitrified brick, is no more than that of hauling a ton of poor limestone or other second-class material the same distance.

Fig. 5.—General view of Wooster pike.

Herewith is a photograph of a section of the Wooster pike as it appears when completed. The cut shows a method of holding the brick in place alongside of the dirt road, instead of using a stone curbing. This plan, devised by me for this purpose, consists of three courses of brick, standing endwise, the first course flush with the top of the pavement, the second breaking joints and dropping 2 inches lower, the third 2 inches lower still, forming a stairwise bond for the brickwork in such a manner that a heavily loaded wagon can not catch and tear up the brick pavement. If a wheel runs off the pavement, it strikes the second course of curbing brick and runs along on that; but it is almost impossible for a wheel to cut through the broken stone filling which surrounds the curbing courses and protects them from the wear of heavily loaded wagons.

The Wooster pike spoken of is herewith shown.

The cost of a road built like the Wooster pike will vary according to location, soil, transportation facilities, etc. This road costs about $16,000 per mile. The same road could be built in this country near a railroad or where transportation facilities are ample for $10,000 or less a mile.

One important feature in the construction of improved roads is the question of grading—whether it is better to give a road the best grade attainable, regardless of an extra cost of from $2,500 to $5,000 per mile, or to do as little grading as possible and be content with a narrower way and steeper grade. These estimates are based upon the cost of main roads running to and from great cities, where it is a necessity to have the best grade possible and all the conveniences that can be afforded at a moderate cost. Of course, for any style of cross roads where the travel is light the system can be modified so that the expense would be accordingly less.

The one particular thing about which people who want a good road must be always careful is to take away the water quickly, and keep it away. One foot deep of macadam material will stand more wear placed upon a road that is thoroughly drained than an unlimited amount if placed upon a road which is unsound—where the soil is full of water and foundation is insecure. My experience goes to show that perfect drainage of the subsoil is necessary in order to construct a good road; and in soils such as the one I have just described it can not be accomplished without putting in drains of a depth of nearly 4 feet. The roads above spoken of, which were made 16 feet wide for heavy travel, were built in a similar manner as to drainage, except that the macadam material was composed of blast furnace slag. The same system of drainage was used in case of the brick road, and so far as I can see the drainage is perfect and the road well adapted for an immense amount of traffic.

The appended diagrams show the depths of macadam, the location of the drains, and the shape of the roadway.

PLAN FOR MACADAM ROAD.

FIG. 6.—Plan for macadam road.

PLAN FOR BRICK ROAD.

FIG. 7.—Plan for brick road.

In addition to the roads built in Cuyahoga County, as described above, there has been a class of roads of a different character, which is destined to be a very great, if not the greatest, factor in our transportation in the future, namely, the electric roads. Notwithstanding the fact that the brick or macadam roads built and being built in Cuyahoga County are the best in the State of Ohio, it remains true that the rate of transportation over these roads with horses and wagons is higher than the rate which prevails upon the electric cars wherever they are introduced.

Side by side the macadam road and the electric road are making their appearance and testing their relative value and importance. The macadam road is built with public money at a cost of $16,000, while the electric road is built with private money at a cost of from $5,000 to $7,000 per mile. Still, the one that is built with private money is furnishing now, and is destined to furnish hereafter, a cheaper means of transportation than can be obtained over the free turnpike with horses and wagons. This being true, the question naturally arises whether it would not be good public policy to still further cheapen the rate of transportation by using the public money to lay down steel rails over which electric cars can go at a greatly reduced cost in transportation, both as to passengers and as to the productions of the farm, rather than stone roads. In this connection I quote from the report of the Ohio Road Commission as follows:

It being the established policy of the people to aid in cheapening transportation by deepening rivers, harbors, and channels, by building roads and bridges, streets and viaducts—all by appropriations of public money and by contributing the use of streets and roads for electric cars—we see no reason why they might not as logically and more profitably contribute to the construction of electric railways to be and remain a part of the common roads, as to the paving of these roads to be operated with horses and wagons, if in any locality the people should desire to do so.

We have already reached the maximum power of horses and other animals for draft, speed, and endurance. The only improvement that we could hope to make to lessen the cost of transportation with these animals would be in improving the roadbed. A comparison of the cost will show that the average expenditure required to macadamize a road or make it hard and durable with any kind of metal is fully equal to the cost required to lay down steel rails, over which not only wagons and carriages propelled by horses, but cars propelled by electric power might also go at a greatly reduced cost in transportation. Gilmore's tables show that the same vehicle can be moved over steel rails with one-eighth of the power that would be required to move it over a macadamized road, and with one-eighteenth of the power that would be required to move it over a gravel road, and with one twenty-fifth of the power that would be required to move it over a common earth road in good condition. Having given the cost of construction, which is the same in each kind of road, and the cost of moving the vehicle being so much less over the steel rail than any other kind of road, and knowing also that the cost of inanimate power is less than the cost of animal power, it seems clear that the substitution of steel rails for macadamized roads, and inanimate power for animal power, is destined to cheapen our transportation in the most effectual manner. The profitable use of steel rails and the application of inanimate power can only be limited by the convenience or inconvenience of terminal facilities, because it will always cost much less to move over a smooth iron rail than over a pavement, whatever power may be applied. This advantage may be neutralized by the disadvantage of loading and unloading, but that is the only thing which, in the long run, will limit the application of this new power.

The Hon. John M. Stahl, in a valuable article in the Illinois number of Good Roads, has made a conservative estimate of the wagon freight of this country for the year 1892 as 500,000,000 tons. He also estimates that this will be transported over the country highways an average distance of 8 miles, which would be equivalent to 4,000,000,000 tons 1 mile. At a cost of 25 cents per ton a mile, which would be required to move it by horse power with ordinary vehicles, it would amount to the enormous sum of $1,000,000,000. This may be stated as the cost of operating wagon roads. Now, if by substituting steel rails and inanimate power, there could be a saving of four-fifths of this amount, which would be much less than the proportion indicated by Gilmore's tables, the cost of moving this tonnage would be only $200,000,000 instead of $1,000,000,000, leaving a gain of $800,000,000. This, for a period of ten years, would leave a net gain of $8,000,000,000.

We have made great unexpected improvements in the means of transportation where we have substituted other power for horse power, while we have made but little improvement in the cost of transportation where we have adhered to animals as the motive power. To this fact should be added another important one, namely, that millions and millions of dollars have been expended to aid in cheapening transportation with horses, while nothing has been expended to aid the means that have been most successful in cheapening our rates of transportation, i. e., electricity. If we should extend the same liberal policy to the electric car that we have extended to horses and wagons by providing a free track for it to go upon, as we have for other vehicles propelled by animals, the rate of transportation would be still further cheapened in the future as it has been in the past, and a lower rate can be reached than by any other means.

It seems probable that the application of electricity to the cars upon our streets and roads is destined to do for the short haul what the steam cars have already done for the long haul. So far as electricity has been applied already, it has shown that the cost of transportation by that means is far less than upon the steam cars, which is indicated by the rate of charge for transportation, the common rate upon steam cars being 3 cents per mile for passengers, while in many cases upon the electric cars it is but 1 cent, or even less, per mile. What has been done by way of cheapening transportation of passengers may be done to a great extent in cheapening the transportation of certain kinds of freight, especially the food products that are raised upon the farms and conveyed to the markets for immediate consumption. This may be done either with or without the aid of public money. Neither the steam cars nor the street cars up to the present time have received the aid of public money; but one element in the cheapening of transportation which has assisted the electric street cars is the use of public streets and roads, which, though not money, is a contribution as valuable as money itself. The cheap rates which prevail upon the street cars could never be attained if the companies that operate these cars were obliged to appropriate and buy the lands upon which they build their roads; so that public aid, though not public money, has been given, and properly given, to cheapen transportation by that means.

The economical advantages are so greatly in favor of steel rails and electric power that no objection can be sustained against their introduction unless it rests upon the supposed inconvenience of using this new means in the most commodious manner. In all our great cities, and most of our smaller ones, double tracks are already laid and are being rapidly extended to the suburbs for considerable distances, from 10 to 15 miles; their use at the present time is entirely confined to the matter of carrying passengers, but after midnight passenger traffic is over, and from that time until 5 o'clock in the morning these tracks are idle and the streets vacant. During that time they could be used to great advantage and with great economy for transporting freight and food products placed upon trail cars to various markets and other places of distribution in the centers of population; so that the question of introducing steel rails and electric power is only a question of extension. The nucleus of the system already exists, and its use can undoubtedly be extended with great advantage.

The effect of this report has been to stimulate the building of electric roads in various parts of the State, all of which have been extremely successful, and have shown the intrinsic value of the electric road to so greatly exceed any other means of transportation for short distances that the public is likely to extend to it the same friendly policy that it has heretofore given to vehicles propelled by animal power; that is to say, the public will provide the track upon which the vehicle runs, while private enterprise will supply the vehicle and power as heretofore. One of the most successful lines yet built through the country, and which is being used for carrying package freight and food products as well as express and mail, is the electric road built from Norwalk to Sandusky via Milan. Cuyahoga County has led the way in the development of electric transportation, but the adjoining counties of Lorain, Summit, and Geauga are fast following her example with increased liberality, and the day is not far distant when every county seat in the State of Ohio will be connected with every other and with every large city by means of a network of electric roads, which should be and probably will be provided by the State cooperating with the various counties. It is not proposed that the public should operate these roads, but only furnish the track, according to the established public policy that has prevailed time out of mind for the public to furnish the way and private enterprise to furnish the power and vehicles.

Taking it all in all, the outlook for a continued development of our road system in Ohio is brighter than ever before. And we shall demonstrate here first, most likely, how far it is wise to provide roads for the wagonload haul by means of horses, and how far it is best to provide steel tracks and inanimate power to supersede the horses.

STONE ROADS IN NEW JERSEY.

By E. G. HARRISON, C. E.,
Secretary New Jersey Road Improvement Association.

PRELIMINARY CONDITIONS.

As New Jersey contains a great variety of soils, there are many conditions to be met with in road construction. The northern part of the State is hilly, where we have clay, soft stone, hard stones, loose stones, quicksand, and marshes. In the eastern part of the State, particularly in the seashore sections, the roads are at their worst in summer in consequence of loose, dry sand, which sometimes drifts like snow. In west New Jersey, which comprises the southern end of the State, there is much loose, soft sand, considerable clay, marshes, and low lands not easily drained.

In addition to the condition of the soil, there is the economic condition must be considered. In the vicinity of large towns or cities, where there is heavy carting by reason of manufactories and produce marketing, it is necessary to have heavy, thick, substantial roads, while in more rural districts and along the seashore, where the travel is principally by light carriages, a lighter roadbed construction is preferred. In rural districts, where the roads are used for immediate neighborhood purposes, an inexpensive road is desirable. The main thoroughfares have to be constructed with a view to considerable increase of travel, as farmers in the outlying districts who formerly devoted their time to grazing of stock, raising of grain, etc., find it more profitable to change the mode of farming to that of truck raising, fruit growing, etc.

The road engineers of New Jersey find that they can not follow old paths and make their roads after one style or pattern. Technical engineering in road construction must yield to the practical, common-sense plan of action. An engineer with plenty of money and material at hand can construct a good road almost anywhere and meet any condition, but with limited resources and a variety of physical conditions he has to "cut the garment to suit the cloth." We start out with this dilemma. We must have better roads, and our means for getting them being very limited, if we can not get them as good as we would like, let us get them as good as we can.

Let me give a practical illustration. Stone-road construction outside of turnpike corporations in West Jersey was begun in the spring of 1891. I was called on by the township committee of Chester Township, Burlington County, to construct some roads. Moorestown is a thriving town of about 3,000 inhabitants in the center of the township. The roads to be constructed, with one exception, ran out of the town to the township limits, being from one-half to 3 miles in length. The roads were generally for local purposes. There were ten roads, aggregating about 11 miles. The bonding of the township was voted upon, and it was necessary, in order to carry the bonding project of $40,000, to have all these roads constructed of stone macadam. The roads to be improved were determined on at a town meeting without consulting an engineer as to the cost, etc., so that the plain question submitted to me was, Can you construct 11 miles of stone road 9 feet wide for $40,000? The conditions to be met were these: There was no stone suitable for road building nearer than from 60 to 80 miles; cost of freight, about 75 cents per ton; the hauls from the railroad siding averaged about 1¾ miles; price of teams in summer, when farmers were busy, about $3.50 per day. In preparation for road construction there were several hills to be cut from 1 to 3 feet; causeways and embankments to be made over wet and swampy ground. For this latter work the property holders and others interested along the road agreed to furnish teams, the township paying for laborers. The next difficulty was the kind of a road to build. As the width was fixed at 9 feet as a part of the conditions for bonding, there seemed only one way left to apply the economics—that was, in the depth of the roads.

On the dry, sandy soils I put the macadam 6 inches deep; this depth was applied to about 6 miles of road. On roads where the heaviest travel would come the roadbed was made 8 inches deep. On soils having springs and on embankments over causeways the depth was 10 inches with stone foundation, known as telford. Where springs existed, they were cut off by underdrains.

It had been the practice of engineers in their specifications to call for the best trap rock for all the stone construction. As this rock is hard to crush and difficult to be transported some 70 or 80 miles to this part of New Jersey, I found that in order to construct all of the road from this best material it would take more money than the bonds would provide; so I had half of the depth which forms the foundation made of good dry sedimentary rock. Of course, in this there is considerable slate, but the breaking is not nearly so costly as the breaking of syenite or Jersey trap rock, and there was a saving of 30 per cent. As the surface of the road had to take all the wear, I required the best trap rock for this purpose.

Since the construction of these roads in Chester Township, roads are now built under the State-aid act by county officials and paid for as follows: One-third by the State, 10 per cent by the adjoining property holders, and the balance (56⅔ per cent) by the county. The roads constructed under this act are generally leading roads and those mostly traversed by heavy teams. They are constructed similarly to those in Chester Township, excepting that they are generally 12 feet wide and from 10 to 12 inches deep. Many of them have a telford foundation, which is now put down at about the same price as macadam, and meets most of the conditions better than macadam. The less expensive stone is used for foundations, and the best and more costly for surface only. In this way the cost of construction has been greatly reduced.

In regard to the width, a road 9 or 10 feet wide has been found to be quite as serviceable as one of greater width, unless it is made 14 feet and over. It is not claimed that a narrow road is just as good as a wide road, but it has been found better to have the cost in length than in width in rural districts. In and near towns, where there is almost constant passing, the road should not be less than from 14 to 20 feet in width. The difficulty in getting on and off the stone road where teams are passing is not so great as is supposed. To meet this difficulty in the past, on each side of the road the specifications require the contractor to make a shoulder of clay, gravel, or other hard earth; this is never less than 3 feet and sometimes 6 to 8 feet in width, according to the kinds of soil the road is composed of and the liability of frequent meeting and passing. In rural districts the top-dressing of these shoulders is taken from the side ditches; grass sods are mixed in when found, and in some cases grass seed is sown. As the stone roadbed takes the travel the grass soon begins to grow, receiving considerable fertilizing material from the washing of the road; and when the sod is once formed the waste material from wear of road is lodged in the grass sod and the shoulder becomes hard and firm, except when the frost is coming out.

Another mode of building a rural road cheaply and still have room for passing without getting off the stone construction is to make the roadbed proper about 10 feet wide, 10 or 12 inches deep; then have wings of macadam on each side 3 feet wide and 5 or 6 inches deep. In case 10 feet is used the two wings would make the stone construction 6 feet wide. If the road is made considerably higher in the center than the sides, as it should be, the travel, particularly the loaded teams, will keep in the center, and the wings will only be used in passing and should last as long as the thicker part of the road.

GRADING.

The preparation of the road and making it suitable for the stone bed is one of the most important parts of road construction. This, once done properly, is permanent. Wherever it is possible the hills should be cut and low places filled, so that the maximum grade will not exceed 5 or 6 feet rise in 100 feet; where hills can not be reduced to this grade without incurring too much expense, the hill, if possible, should be avoided by relaying the road in another place.

Wherever stone roads have been constructed it has been found that those using them for drawing heavy loads will increase the capacity of their wagons so as to carry three or four times the load formerly carried. This can easily be done where the road has a maximum grade of not greater than 5 or 6 per cent, as before stated; but when the grade is greater than this the power to be expended on such loads upon such grades will exhaust and wear out the horses; thus a supposed saving in heavy loading may prove to be a loss.

In the preparation of the road it is necessary to have the ditches wide and deep enough to carry all the water to the nearest natural water way. These ditches should at all times be kept clear of weeds and trash, so that the water will not be retained in pools. Bad roads often occur because this important matter is overlooked.

On hills the slope or side grade in construction from center of road to side ditches should be increased so as to exceed that of the longitudinal grade; that is, if the latter is, say, 5 per cent, the slope to side should be at least 6 per cent and over.

Where the road in rural districts is on rolling ground and hills do not exceed 3 or 4 per cent, it is an unnecessary expense to cut the small ones, but all short rises should be cut and small depressions filled. A rolling road is not objectionable, and besides there is no better road-bed for laying on metal than the hard crust formed by ordinary travel. In putting on the metal, particularly on narrow roads, the roadbed should be "set high;" it will soon get "flat enough." It is better to put the shouldering up to the stone than to dig a trench to put the stone in. If the road after preparation is about level from side to side and the stone or metal construction is to be, say, 10 inches deep, the sides of the roadbed to receive the metal should be cut about 3 inches and placed on the side to help form the shoulder; the rest of the shoulder, when suitable, being taken from the ditches and sides in forming the proper slope. The foundation to receive the metal, if the natural roadbed is not used and the bed is of soft earth, should be rolled until it is hard and compact. It should also conform to the same slope as the road when finished from center to sides. If the bed or foundation is of soft sand rolling will be of little use. In this case care must be taken to keep the bed as uniform as possible while the stone is being placed on the foundation.

When the road passes through villages and towns the grading should reduce the roadbed to a grade as nearly level as possible. It must be borne in mind that the side ditches need not necessarily always conform to the center grade of the road. When the center grade is level the side ditches should be graded to carry off the water. In some cases I have found it necessary to run the grade for the side ditches in an opposite direction from the grade of the road. This, however, does not often occur. The main thing is to get the water off the road as soon as possible after it falls, and then not allow it to remain in the ditches. And just here the engineer will meet with many difficulties. The landowners in rural districts are opposed to having the water from the roads let onto their lands, and disputes often arise as to where the natural water way is located. This should be determined by the people in the neighborhood, or by the local authorities. I have found in several cases, where the water from side ditches was allowed to run on the land, that the land was generally benefited by having the soil enriched by the fertilizing matter from the road.

METAL CONSTRUCTION.

After the roadbed has been thoroughly prepared, if made of loam or clay, it should be rolled and made as hard and compact as possible. Wherever a depression appears it should be filled up and made uniformly hard. Place upon it a light coat of loam or fine clay, which will act as a binder. If the roller used is not too heavy it may be rolled to advantage, but the rolling of this course depends upon the character of the stones. If the stones are cubical in form rolling is beneficial, but if they are of shale and many of them thin and flat, rolling has a tendency to bring the flat sides to the surface. When this is the case the next course of fine stone for the surface will not firmly compact and unite with them.

When the foundation is of telford it is important that stones not too large should be used. They should not exceed 10 inches in length, 6 inches on one side, which is laid next to the earth, and 4 inches on top, the depth depending on the thickness of the road. If the thickness of the finished road is 8 inches the telford pavement should not exceed 5 inches; if it is 10 or more inches deep, then the telford could be 6 inches. It need in no case be greater than this, as this is sufficient to form the base or foundation of the metal construction. The surface of the telford pavement should be as uniform as possible, all projecting points broken off, and interstices filled in with small stone. Care should be taken to keep the stone set up perpendicular with the roadbed and set lengthwise across the road with joints broken. This foundation should be well hammered down with sledge hammers and made hard and compact. Upon this feature greatly depends the smoothness of the surface of the road and uniform wear. If put down compactly rolling is not necessary, and if not put down solid rolling might do it damage in causing the large stones to lean and set on their edges instead of on the flat sides. I refer to instances where the road is to be 10 inches and over. Then put on a light coat or course of 1½-inch stone, with a light coat of binding, and then put on the roller, thus setting the finer stone well with the foundation and compacting the whole mass together.

SURFACE STONE.

After the macadam or telford foundation is well laid and compacted the surface, or wearing stone, is put on. If the thickness of the road is great enough, say 12 or 14 inches, this surface stone should be put on in courses, say of 3 and 4 inches, as may be required for the determined thickness of the road. On each course there should be applied a binding, but only sufficient to bind the metal together or fill up the small interstices. It must be remembered that broken stone is used in order to form a compact mass. The sides of the stone should come together and not be kept apart by what we call binding material; therefore only such quantity should be used as will fill up the small interstices made by reason of the irregularity of the stone. Each course should be thoroughly rolled to get the metal as compact as possible. When the stone construction is made to the required depth or thickness, the whole surface should be subjected to a coat of screenings about 1 inch thick. This must be kept damp by sprinkling, and thoroughly rolled until the whole mass becomes consolidated and the surface smooth and uniform. Before the rolling is finished the shoulders should be made up and covered with gravel or other hard earth and dressed off to the side ditches. When practicable these should have the same grade or slope as the stone construction. This finish should also be rolled and made uniform, so that, in order that the water may pass off freely, there will be no obstruction between the stone roadbed and side ditches. To prevent washes and insure as much hardness as possible on roads in rural districts grass should be encouraged to grow so as to make a stiff sod.

MATERIAL.

For shouldering, when the natural soil is of soft sand, a stiff clay is desirable. When the natural soil is of clay, then gravel or coarse sand can be used, covering the whole with the ditch scrapings or other fertilizing material, where grass sod is desirable. Of course this is not desirable in villages and towns.

For binding, what is called garden loam is the best. When this can not be found use any soft clay or earth free from clods or round stones. It must be spread on very lightly and uniformly.

Any good dry stone not liable to disintegrate can be used as metal for foundation for either telford or macadam construction. For the surface it is necessary to have the best stone obtainable. Like the edge of a tool, it does the service and must take the wear. As in the tool it pays to have the best of steel, so on the road, which is subject to the wear and tear of steel horseshoes and heavy iron tires, it is found the cheapest to have the best of stone.

It is difficult to describe the kind of stone that is best. The best is generally syenite trap rock, but this term does not give any definite idea. The kind used in New Jersey is called by the general name of Jersey trap rock. It is a gray syenite, and is found in great quantities in a range running from Jersey City, on the Hudson River, to a point on the Delaware River between Trenton and Lambertville. There are quantities of good stone lying north of this ledge, but none south of it.

The best is at or near Jersey City. The same kind of stone is found in the same ranges of hills in Pennsylvania, but in the general run it is not so good. The liability to softness and disintegration increases after leaving the eastern part of New Jersey, and while good stone may be found, the veins of poorer stone increase as we go south and west.

It is generally believed that the hardest stones are best for road purposes, but this is not the case. The hard quartz will crush under the wheels of a heavy load. It is toughness in the stone that is necessary; therefore a mixed stone, like syenite, is the best. This wears smooth, as the rough edges of the stone come in contact with the wheels. It requires good judgment based on experience to determine the right kind of stone to take the constant wear of horseshoes and wagon tires.

REPAIRS.

If good roads are desired, the work is not done when the road is completed and ready for travel. There are many causes which make repairing necessary. I will refer to only a few of them. Stone roads are liable to get out of order because of too much water or want of water; also, when the natural roadbed is soft and springy and has not been sufficiently drained; when water is allowed to stand in ditches and form pools along the road, and when the "open winters" give us a superabundance of wet. Before the road becomes thoroughly consolidated by travel it is liable to become soft and stones get loose and move under the wheels of the heavily loaded wagons. In the earth foundation on which the stone bed rests the water finds the soft spots. The wheels of the loaded teams form ruts, and particularly where narrow tires are used.

The work of repair should begin as soon as defects appear, for, if neglected, after every rain the depressions make little pools of water and hold it like a basin. In every case this water softens the material, and the wagon tires and horseshoes churn up the bottoms of the basins. This is the beginning of the work of destruction. If allowed to go on the road becomes rough, and the wear and tear of the horses and wagons are increased. Stone roads out of repair, like any common road in similar condition, will be found expensive to those who use and maintain them. The way to do is to look over a road after a rain, when the depressions and basins will show themselves. Whenever one is large enough to receive a shovelful of broken stone, scrape out the soft dirt and let it form a ring around the depression. Fill with broken stone to about an inch or two above the surface of the road. The ring of dirt around will keep the stone above the surface in place, and the passing wheels will work it on the broken stone and also act as a binder. The whole will work down and become compact and even with the road surface. The ruts are treated in the same way. Use 1½-inch stone for this; smaller stones will soon grind up and the hole appear again.

The second cause of the necessity for road repairs is want of water. This occurs in summer during hot, dry spells. The surface stone "unravels;" that is, becomes loose where the horses travel. This condition is more liable to be found on dry, sandy soils, and where the roadbed is subject to the direct rays of the sun, and where the winds sweep off all the binding material from the surface. In clay soil there is little or no trouble from "unraveling." The cause being found, the remedy is applied in this way: Put on water with the sprinkler before all the binding material is blown off. If the hot, dry weather continues sprinkling should continue. Do this in the evening or late in the afternoon.

The next mode is to repair the road by placing the material back as it was originally. The loose stones are placed in the depressions and good binding material—garden loam or fine clay—is put on, then roll the whole repeatedly and dampen by sprinkling as needed until the whole surface becomes smooth and hard. Care must be taken that too much binding material is not used. If too much is used it will injure the road in winter when there is an excess of water.

When a road has been neglected and allowed to become uneven and rough, or is by constant use worn down to the foundation stones, there should be a general repairing. In the first place, if it is the roughness and unevenness that is the only defect, this may be remedied by the use of a large, heavy roller with steel spikes in its rolling wheels. This will puncture the surface so that an ordinary harrow will tear up the surface stones. Then take the spikes out of the roller wheels, and, with sprinkling and rolling, the roadbed can be repaired and made like a new road. But if the cause of the roughness is from wearing away of the stone, so that the surface of the road is brought down to or near the foundation, then the road needs resurfacing. The mode of treatment is the same as in the other case.

NECESSARY MACHINERY.

In districts where there is stone suitable for road construction the county, town, township, or other municipality, proposing to construct stone roads, should own a stone quarry and a stone crusher. For grading and preparing the road for construction, dressing up sides, clearing out side ditches, etc., a good road machine is necessary. For constructing roads and repairing them a roller is necessary, the weight depending upon the kind of road constructed. If the road is not wide a roller of from 4 to 6 tons is all the weight necessary. The rolling should be continued until compactness is obtained. For wide, heavy roads a steam roller of 15 tons can be used to advantage. A sprinkling wagon completes the list that is necessary for the county or town or other municipality constructing its own roads.

MACADAMIZED ROADS.
By CHARLES E. ASHBURNER, Jr., C. E.

Roads of broken stone have been the subject of violent disputes for many years. The chief point on which the best authorities differ is the question of a paved foundation; such a foundation being favored by the Telfordites (if one may coin the word), and deemed unnecessary by the followers of Macadam.

There is no doubt that with the use of good judgment and care in building stone roads the expensive hand-set pavement required by Telford can be dispensed with. This opinion is based upon the practical experience of many years and upon the results of many careful experiments. The chief objections to the telford system are, briefly, these: (1) The original cost, which is about one and one-half times that of the macadam system; (2) when a telford road begins to wear, it requires entirely new "surfacing," since the large, hand-set stones protrude, causing a very rough road. The cost of resurfacing is nearly equal to the original cost of a macadam road.

Believing that the future roads in the United States must be the same as those built by John L. Macadam in England about 1816, the writer will confine himself to this branch of road building exclusively.

The first thing the engineer should consider is the location, and then the drainage; but as these very important points in construction have been so thoroughly written up, it is only necessary to call the attention of road builders to these most essential factors and pass on to other important considerations.

PREPARATION OF THE ROADBED.

Next comes the preparation of the roadbed. This should be well rolled, if practicable, by a steam roller weighing at least 10 tons. The great advantages of a steam roller over one drawn by horsepower are: (1) The reduction of cost; (2) the extra weight, which consolidates the stone before the continued passing to and fro has worn off the sharp angles; (3) the fact that the stone is not disturbed by the horse's feet, as in the case of a horse roller.

Every hole should be carefully filled with the same material that composes the rest of the roadbed; and the finished cross section of the roadbed should be formed by two grades of 1 in 30 from the sides, uniting at the center, with the apex slightly rounded.

A roadbed can almost always be formed on these lines; but there have been cases where it was impossible, even after carefully tile-draining the roadbed and allowing it to stand for some time, to get a steam roller weighing 10 tons upon the ground. In such case it is recommended that the roadbed be excavated for 6 inches below its finished grade. This excavation should be filled with crushed stone of about $2\frac{1}{2}$ inches diameter, over which should be spread 2 inches of coarse sand. Roll with a light roller, weighing from 1 to 2 tons, or simply allow traffic to pass over the same. In a short time the stone will become so "bonded" as to enable work to proceed on the road. By adopting this plan, not forgetting the tile-drain, roads have been built over places previously too soft to bear a man's weight.

THE BEST ROADBED.

The best road is that of which the particles composing it are as nearly as possible in their natural condition. To obtain such at a minimum cost should be the aim of the engineer. Spread 4 inches of crushed stone on a carefully prepared roadbed. The largest cubes of this stone should pass with ease through a 2-inch ring, and the smallest be just too large to pass through a stationary screen of $1\frac{1}{4}$-inch mesh, inclined at an angle of 45 degrees; that is, about $\frac{1}{2}$-inch diameter. A greater thickness than 4 inches will cause the stones to "elbow" one against the other, thereby wearing off their sharp corners, making "bonding" a very slow process, if not an impossibility. Sprinkle, making the stone as wet as possible without softening the roadbed; then pass the roller a few times until the stone is only slightly disturbed by teams returning with empty carts. Now apply a second coat, but 3 inches thick, of similar stone. Sprinkle and roll a second time, until carts make no impression upon the stone. When the stone has been bonded to this extent, and while thoroughly wet, apply a thin coating of screenings in size from one-half inch to dust. Do not dump in piles and spread, but take the screenings from the cart in shovels and "whisk" over the surface just thick enough to fill small spaces between stones. Do not sprinkle after putting on the "binder," for by so doing you would cause the wheels of the roller to take up fine particles, to the detriment of the road.

Of course, where traffic is very heavy or the winter very severe, two coatings of 4 inches each are recommended, and then one of 3 inches. But in Virginia and other Southern States a well-constructed macadam road 7 inches thick with a very moderate amount of care will last from six to eight years before it will require a general resurfacing. This does not apply to roads built of limestone, but to those built of granites and trap rocks.

After leaving the roadbed never allow anything but road metal to go into the road. The practice of using clay or any other earth as a binder is absolutely unnecessary and wrong; unnecessary, because, if the stone is broken into cubes from one-half inch to 2 inches in diameter and well rolled, it will form an almost indestructible bond; wrong, because anything of an earthy nature has a tendency to retain the water which falls on the surface, and in a short time to cause the road to break up, especially in a climate where freezing and thawing alternate.

Sand, which is recommended by many engineers, does not make as good a binder as granite screenings or screenings of the material of which the road is built. A road on which sand has been used never has quite the same compact surface as one built wholly of one material. It has a tendency to shake to pieces in very dry weather. These remarks apply also to gravel as a "binder."

I have tried building roads with unscreened stone—that is, fine and coarse as it comes from the crusher—ranging in size from 2 inches in diameter to dust; but this does not make as smooth a road, and consequently not so lasting, as one built of screened stone. The finest stone jolts to the bottom of the cart in transportation, and when spread is very apt to form in a hard knot which does not roll uniformly with the larger stone.

MAINTENANCE OF A ROAD.

The old saying, "A stitch in time saves nine," never applied more appropriately to anything than it does to the maintenance of a macadam road.

Inspect your roads constantly and carefully; never allow the smallest hole to remain, but use the pick to loosen the surface as one forms and then carefully fill with chips one-half inch in diameter, or even smaller, of the same material of which the road is built, and roll. In filling be careful not to change a hole into a hillock, which would eventually cause two holes, one on each side. Equal attention should be paid to maintaining thorough drainage, so that the water will run off without saturating the edges of the road. When the road surface at last becomes worn out, pick it thoroughly (picking by the steam roller is by far the most economical), then apply stone, and proceed as in the original construction.

Roads now in my charge, built four years ago of Virginia gray granite (rejecting such as contained much mica), were only 7 inches thick. They have been constantly under heavy traffic of the worst kind, namely, country teams, which drive one behind the other in the center of the road, yet not one cent has been spent in repair, and they are as free from holes as the day they were constructed. They are worn, however, as the fine granite dust taken from the gutters will prove; but the wear has been denudation simply, owing to the fact that they were constructed upon a roadbed of uniform hardness and smoothness, and all material used was uniformly tough.

AN INSTANCE OF ROAD ECONOMY.

A very striking example of the economy of building macadamized roads came under my observation recently. A machine weighing 16,000 pounds was drawn 4 miles on the Brook turnpike, a macadamized road. It required four mules (4,000 pounds to a mule) and one and one-half hours of time, at a cost of 15 cents per mule per hour, or a total cost for 4 miles of 90 cents. After traveling 4 miles on the macadamized turnpike the route lay a little less than 2,000 feet on a dirt road. To travel this 2,000 feet it was necessary to use ten of the best mules and seven men; and with this force it took nine hours to complete the journey. The cost was $19.80, at which rate 4 miles would have cost $209.08; or, in other words, $208.18 absolutely thrown away for want of a macadam road. A macadam road, such as would have prevented this enormous waste of money, would cost about $100 per mile for every foot of width; that is, a 12-foot road, $1,200 per mile; a 16-foot road, $1,600 per mile, etc. One can well realize from this the enormous sum wasted annually by our present impassable highways.

ROADS.

By FRANCIS V. GREENE,
President of the Barber Asphalt Paving Company.

[Address before the students of Union College.]

PRESIDENT RAYMOND AND STUDENTS OF UNION COLLEGE: My predecessors in this course of lectures have not only been men of distinction, but they have also spoken to you on very distinguished subjects. You have heard about diplomacy from the ambassador to Great Britain; about journalism from the foremost editor of the day; about wealth from one of the most successful accumulators of that desirable article, and you have heard from others specially qualified to speak about other subjects which appeal to the imagination. Compared with these the subject of roads is essentially commonplace, a part of the everyday routine of the struggle for existence, and yet it is true that roads and civilization go hand in hand, each mutually dependent on the other, and each in turn cause and effect of the other. The transition from the nomadic, pastoral, semibarbarous manner of life to a commercial, trading, militant, organized civilization is marked by the construction of roads. Considering roads in the broadest sense as means of communication and transport on land, and thus including railroads, common roads, and city streets, it is true now, as it has been for three thousand years, that the degree of civilization to which any people have attained is accurately measured and indicated by the condition of their roads. The history of Rome, of ancient India, and of Peru amply attest the accuracy of this statement. But roads belong to the material side of civilization; they form part of the broad and solid foundation upon which the structure is reared, and of which literature, philosophy, science, and the arts are the domes and pinnacles.

Of this essentially prosaic subject I am to speak to-day in the effort to show you, on the one hand, what relation roads bear to civilization and the prosperity and comfort of mankind, and, on the other hand, how they should be constructed and maintained.

At the outset you will naturally ask how it is, if roads are so intimately connected with civilization, that the United States, which claims to be among the most civilized nations in the world, should confessedly have roads so bad that they are justly described as intolerable. But the answer is not far to seek. The United States have the longest and best roads in the world. But they are in the form of railroads. And the construction of these railroads has absorbed so much energy and capital that there has not until now been time to construct good common roads, nor has the necessity for them been evident.

It is well to glance briefly at the origin and development of this magnificent system of railroads. When the colonies from which the United States have grown were first settled, they occupied a narrow strip along the Atlantic seacoast, and such settlements as were made in the interior were made on the banks of navigable rivers. Such commerce as they had, not only with Europe, but among themselves, was carried on by means of vessels. This condition continued down to the Revolution, and for nearly a generation after it.

When during the first third of this century the population began to spread westward the necessity for good roads became manifest, and during the time of Monroe, Clay, and Calhoun the question of internal improvements was one of the burning questions of national politics; those who believed in a liberal construction of the Constitution being favorable to the construction of roads by the General Government, and those who insisted on a strict construction of the Constitution denying the power of the General Government to spend money for any such purpose. It was finally decided that the General Government should undertake the construction of a national road, which, following up the valley of the Potomac, should cross the Alleghanies, descend to the Ohio at Wheeling, and then go on to St. Louis. This work was begun in 1806, but it was carried on very slowly, and before much of it had been finished steam railroads were introduced, and it was seen at once that they would be immensely superior to the old form of road. The construction of the national road was therefore abandoned, and private capital undertook the construction of steam railroads; and for sixty years this has continued, until now the United States has nearly 180,000 miles of line, and more than 50,000 miles more of second track and sidings. In this work the best talent, both in a mechanical and administrative capacity, has been employed for the last two generations. The result is that while the United States has but one-twentieth of the population of the world it has more than one-half of its railroads; it has but one-fifth the population of Europe, but it has one and a half times as many railroads.

These railroads are in a large measure the source of its rapid growth in wealth, for they have penetrated wherever the population has penetrated and often in advance of it, thus affording facilities for communication and for commerce such as are possessed by no other country in the world. The capital invested in them is counted by the thousands of millions of dollars, and their employees by the hundreds of thousands. The task has been so prodigious that the amount of either capital or thought that could be devoted to the construction of other forms of communication has been comparatively small.

It is evident, however, that there must be a limit to the building of railroads, and it would seem as if that limit had been practically reached in certain parts of the country. It is impossible to have a railroad leading to every farm, although this condition is closely approximated in New Jersey, where, it is said, there is no point in the State which is more than 7 miles from a railroad. In the older and more settled portions of the country the railroads are so numerous and the rates are so low that they yield but a small return on the capital invested, and the construction of new railroads has ceased to be an attractive field for investment. The rates of freight have been steadily reduced year by year until they are now barely one-fourth of what they were thirty years ago. Still the transportation problem can not be considered as satisfactorily solved if it costs as much to carry a ton of wheat or potatoes to the railway station as it does to carry it 400 miles over the railroad. So that with the practical completion of the railway system in a large section of the country an agitation has sprung up in favor of the improvement of the common roads. These are needed for a double purpose; first, for the purposes of commerce to serve as feeders to and distributors from the railroads; second, for the purpose of health and pleasure, for the use of bicycles and pleasure carriages.

As already stated, the common roads have been comparatively neglected during the construction of the railroads. Still we have an enormous number of roads; in bad order for the most part. In New York there are 8,110 miles of railroad and about 80,000 miles of common road; in Massachusetts, 2,121 miles of railroad and 17,145 miles of common road; in New Jersey, 2,176 miles of railroad and about 18,000 miles of common road. Statistics for all the States are not available, but if the ratio were the same in them as in the three States named the total length of roads in the United States would be about 1,800,000 miles. It is probably not so great as this. Gen. Roy Stone estimates it at something over 1,300,000 miles. These roads have grown up regardless of system or method, and for the most part have been built without reference to engineering principles. As the country was settled rude tracks were laid out connecting neighboring villages, and these in turn were united to form highways between towns and villages. The method of constructing and maintaining them was by the "labor tax," in which each taxpayer was required to furnish a certain number of days' labor in the spring for the purpose of digging up the ditches and throwing their contents in the middle of the road. Their condition was so bad that about fifty years ago private capital was invoked for the purpose of improving the more important roads, and charters were given in many States for "turnpike companies," which were authorized to collect toll from every passing vehicle, animal, and man, and in return were required to keep the road in order. The system was never a success; the farmers considered the tax unjust, the roads were not kept in proper order, and many of the companies lost money. Some surrendered their charters and others were bought out by the State or county. The old system of the "labor tax" was then restored, or in lieu of it a money tax was levied. Until within the last few years this system was universally followed, each county taking care of its own roads, and by means of a road tax in the form of either labor or money. The State exercised no supervision, and skilled engineers as a rule were never employed. In Massachusetts, the road expenditures, outside of cities, in 1893 were $1,136,944. or $66.30 per mile; in New Jersey, $778,470.82, or $43.24 per mile; in New York, about $2,500,000 or about $30 per mile. If the average expenditure in other States was only $18 per mile the total for the entire country would be about $20,000,000. It is not too much to say that the greater part of this sum produced no useful result, and was wholly wasted.

The bad condition of the roads began to attract widespread attention something over ten years ago. Certain elementary principles were evident at a glance, to wit: The price of farm products is fixed at the great cities or centers of consumption and distribution, and is wholly beyond the farmer's control, and the cost of transportation is a principal factor in determining his profits or the possibility of any profit. On the railroads this has been reduced until it varies, according to bulk, from 1 cent to 6 mills per ton per mile. But the average roads are so bad that a two-horse team and wagon, the value of which is $3 per day, can not haul a ton of produce more than ten miles and return in a day. The cost of road transportation is, therefore, 30 cents per ton per mile, or about forty times as great as the rate on the railroad. The average distance from the farm to the nearest railway station is at least ten miles, so that it costs as much to get the goods to or from the railroad station as to carry them 400 miles on the cars. It only needs to state these elementary facts to show what an enormous drain bad roads make on our resources.

It is evident that an improvement in these conditions is imperative, and the remedy is equally evident, for it has been proved not only by mechanical experiment, but by actual test, that the same force which draws 1 ton on a muddy earth road will draw 4 tons on a hard macadam road. On the improved roads in New Jersey loads of 4 to 5 tons are habitually drawn by a two-horse team. This effects a saving of fully three-fourths of the cost of hauling to the station, and reduces the cost of road transportation from 30 cents to 7½ cents per ton per mile. What this saving amounts to may be imagined when it is known that the New York Central Railroad alone carries nearly 20,000,000 tons of way freight in a year. If this is hauled only 2 miles by road to or from the station, and a saving of 22½ cents per ton per mile could be effected, it would mean a total saving of nearly $9,000,000. These figures may seem exaggerated, but they will no longer appear so when we realize the saving actually accomplished by the reduction in railroad rates in the last twenty-five years. For instance, in 1869 the average freight rate on the New York Central Railroad was 2$\frac{1}{10}$ cents per ton mile; in 1893 it was 7 mills. This saving, on the business of 1893, is upward of $61,000,000! This is the result which has been accomplished by the application to the railroad problem of the highest available talent. During these same twenty-five years little or no attention has been given to the road problem. The roads are as bad now as they were in 1869, and the cost of transportation over them is as great now as it was then. In the next twenty-five years the results accomplished on the common roads are likely to be as remarkable as those achieved on the railroads in the last twenty-five years.

So much for the dollars-and-cents side of the road-improvement question. But there is another and hardly less important side, and that is the use of the roads for health and pleasure; and this appeals not so much to the farmer as to the inhabitants of cities. At the beginning of this century 3 per cent of the population lived in cities of 8,000 inhabitants or upward, and there were six such cities; in 1890 there were 448 such cities, and about 30 per cent of the entire population lived in them. In New York about 60 per cent of the population lives in cities, and in Massachusetts 69 per cent. In proportion as the urban population grows, and possibly in still greater proportion, the number increases of those who desire to escape to the country for pleasure during a part of the year; and nearly all country pleasures, sports, and amusements are dependent in a greater or less degree on the condition of the roads. The driving element, not only in the more expensive form of coaching, but also in the plain American buggy and carryall, is constantly increasing. But the most extraordinary increase among those who find pleasure on the road is in the number of cyclists or wheelmen, and lately wheelwomen. Colonel Pope, who is one of the pioneers in the manufacture of this machine, and at the same time a most ardent and effective advocate of good roads, estimates the number of bicycles made in the last two years at 750,000, and the number that will be in use in 1895 at close to 1,000,000.

Every one of these wheelmen is a preacher, in season and out of season, of the gospel of good roads; and they are not scattered and disunited like the farmers, but they live in cities, are thoroughly organized, have their clubs and leagues, support a monthly magazine with a circulation of 100,000 copies devoted to Good Roads and called by that name, have a chief consul at their head, make their wishes known with no uncertain sound in legislative halls, and at the polls are disposed to consider all political issues secondary in importance to that of road improvement. Their influence in the agitation for good roads has been of the highest value, and it is quite probable that they will be more potent in framing road legislation than any other class. A machine which enables a man to travel with pleasure, without discomfort, and practically without expense, 40 miles in a day, is evidently one which "has come to stay," and the number of wheelmen is likely to reach extraordinary proportions in the next few years. And, as already stated, every one of them is a preacher of road improvement.

The agitation for good roads thus rests on two distinct bases, business, or economy in transportation, and pleasure. It has been in progress with ever-increasing volume for more than ten years. During that time, as General Stone has shown, "sixteen States have passed new road laws, more or less radical in their nature, and one has amended its constitution to permit the adoption of such laws. Many hundreds of miles of good roads have already been built, in localities widely separated, under varying conditions, and through various methods of administration, finance, and construction." It has also been proposed to have roads constructed by the Federal Government, but this idea has met with little encouragement. Out of it, however, has grown a law passed by Congress in 1893 providing for a bureau in the Department of Agriculture to collect information in regard to road improvement in the different States, and to disseminate this by means of publications. The bureau consists simply of a special agent, Gen. Roy Stone in charge, and his clerks; but the information which it has obtained and published is of great value. His inquiries have been confined to, first, the legislation passed or proposed in the different States; second, the methods and cost of construction in different localities; third, the existence of suitable road materials in different parts of the country, and the rates at which railroads are willing to transport them. The bureau has now published ten bulletins in pamphlet form, and they contain information of great value which is not elsewhere accessible.

The general trend of the legislation enacted in the sixteen States before referred to is to provide that the road tax shall be paid in money and not in labor, to authorize the county supervisors under certain conditions to assume entire charge of the roads, and to issue county bonds for their improvement. But to this rule there have been important exceptions. In Pennsylvania an act is pending which requires the State to pay $1,000,000 per annum, this sum to be divided among the counties in proportion to the road tax paid in each. In Massachusetts a State highway commission has been appointed, which has already collected a mass of important statistics, a knowledge of which is necessary to a proper study of the problem; this commission is authorized under certain conditions to assume control of any particular highway, designate it as a State road, and improve and maintain it at the expense of the State, subject to appropriations made by the legislature. In New York two measures have been proposed and have passed one branch of the legislature, but failed to become laws. One contemplated the construction of State highways connecting the adjacent county seats, to be paid for by the State at an estimated cost of $10,000,000. The other provided for the construction or improvement of roads under the supervision of State and county officials, one-third of the cost to be paid by the State, one-third by the county, and one-third to be assessed upon the adjacent property.

But the State in which the most novel legislation has actually been enacted and in which the most important practical results have been obtained is New Jersey, and it is worth while to examine these laws and the effect of them somewhat in detail. The first law, passed in 1888, and enlarged in 1891, abolished the road overseers, gave the township committee full control over the roads in the township, authorized them to levy taxes and borrow money for road improvement, and required all road taxes to be paid in money. The second law, passed in 1889 and amended in 1891, authorized the county freeholders, on a vote of a majority of the voters in the county, to assume exclusive control of any road in the county, to levy taxes and borrow money for its improvement, and to assess one-third of the cost on the cities or townships in the county and the remaining two-thirds on the county at large. The act further provided for letting the work by contract on definite plans and specifications, and under the supervision of a competent engineer. Under these laws Essex County, "though only 12 miles square, has built more than 200 miles of fine telford and macadam roads; * * * Union County has borrowed $435,000 at 4 per cent on 5-20 bonds and covered the county with a complete system of telford and macadam roads; * * * and Passaic County, adjoining Essex and Union, has built during the past four years about 65 miles of macadam roads." Moreover, and most important, "with the interest of the bonds added to the annual tax levy, the rate of taxation is lower than before the building of the roads," and the value of the property along the roads has increased 30 to 50 per cent. These quotations are from the annual report of Edward Burrough, State commissioner of public roads.

The third and most radical of the laws is what is known as the State-aid law, passed in 1891. It provides that on petition of the owners of two-thirds of the lands bordering on any public road not less than 1 mile in length, praying that the road may be improved, and agreeing to pay one-tenth of the cost, the county freeholders shall improve the road, and one-tenth of the cost shall be paid by the abutting property, one-third by the State, and the balance (56⅔ per cent) by the county. The State is limited to an expenditure of $75,000 in any one year for its share of such improvements, and the county to one-half of 1 per cent of its assessed valuation. Under this law 10 miles of road were built in 1892, 25 miles in 1893, over 60 miles in 1894, and a still larger amount is projected for 1895; the applications being in excess of the limit named in the law. At first the cost was about $6,000 per mile, but by reducing the width of the metaled part of the road, and by a large reduction in the price of material, the cost has been reduced to $3,000 per mile. At this rate the law makes possible an expenditure of $225,000 a year, which will build 75 miles of roads.

These laws appear to afford a satisfactory solution of the problem. They provide the machinery for improving the roads in any one of three methods, at the expense of the township, at the expense of the county, or a division of the expense in certain proportions between the State, the county, and the abutting property, and the decision as to the method is determined by the votes of those most interested. These laws have produced more definite results than those in all the remaining fifteen States combined, and they have made the roads in New Jersey famous throughout the land. They are popular with all classes of the people in that State, and they are worthy of careful consideration by the legislators in other States. In States like New York and Massachusetts, where two-thirds of the population and three-fourths of the assessed valuation of property are in the cities, the provision for State aid enables and requires the cities to pay a share of the cost, and this is manifestly proper, since they share largely in the benefits. The cost to the farmer, who derives the greatest benefit, is reduced to a bagatelle. General Stone states that in New Jersey the annual road tax is about 10 cents per acre and the assessment about 4 cents additional. In spite of this small cost it is a remarkable fact that the road laws in New York providing for improvement at the cost of the State, under which three-fourths of the expense would fall upon the cities, have been defeated by the representatives of the farmers. Possibly when the matter is more fully understood the result will be different.

And now in regard to the construction of roads. All historical accounts of roads begin with the famous Roman roads. Wherever the Roman armies penetrated, in Africa, in Thrace, in Spain, in Gaul, and even in Britain, they spent a considerable part of their time in building solid roads, and many of them are to be seen to this day. In France hundreds of miles of them serve as the foundations of the existing roads of that country. The Roman roads were about 3 feet thick and consisted of four layers; first, a layer of large stones laid dry; second, a layer of rubble masonry or coarse concrete; third, a layer of fine concrete; fourth, a layer of dressed stone or paving blocks. These roads were solid and durable, and their lines were well laid out, but in no other respects were they good. They were at least three times too thick, involving a useless expenditure of labor and materials, which is the most unpardonable fault in engineering constructions. And they were intolerably rough, especially as the Romans had no springs on their vehicles. During the middle ages the roads were everywhere neglected. The art of road building was first revived in France in the seventeenth century, and in the eighteenth century it made great progress under a celebrated engineer named Tresaguet, who anticipated by two generations the method of Telford. An enormous amount of road building or rebuilding was done under Napoleon during the consulate and the Empire, and the admirable system of French roads, which are generally considered the finest in the world, was then substantially completed.

For the last eighty years the efforts of the French engineers of the Ponts et Chaussées have been devoted to the maintenance of existing roads rather than the construction of new ones. To this subject of maintenance the best engineers in France have devoted their thought and study; they have written numbers of text-books and memoirs on the subject, and they have brought the art to the highest point of perfection. In England the roads remained in a horrible condition until the early part of this century, when they were nearly all rebuilt by Telford and Macadam. Telford was an educated engineer and architect, who, in addition to roads, built houses, docks, canals, and famous bridges like that over the Menai Straits. He was engaged in reconstructing the roads of Scotland, England, and Wales at various times from 1803 to his death in 1834. Macadam had not the benefit of a scientific education, but prided himself on being a practical road master, and not an engineer. He was appointed superintendent of the roads in the Bristol district in 1816, and between that date and his death, in 1837, he rebuilt over 20,000 miles of roads in various parts of Great Britain.

I do not deem it necessary to go into details in regard to the roads of Telford and Macadam, for the reason that equally good roads have recently been built in America, of which I shall speak later on. Both built their roads of broken stone and drained them thoroughly. Telford used a foundation of larger stones placed in position like a rough stone pavement, and he used fine material to bind the surface. Macadam discarded the foundation, or "pitching," rejected the use of binding material, and insisted on having the road compacted and its smooth surface formed by the action of the vehicles after the road was opened to traffic—a painful and tedious process.

The subject of the construction of roads is a very large one; it has occupied the attention of engineers for several generations, and a great number of books has been written in regard to it. There is also a great variety in the forms of construction, depending upon the traffic to be carried, the nature of the country over which the road is to pass, the road materials available, and the amount of money which can be used for construction. It is impossible in a brief discussion like this to go into the matter at all in detail, and I do not think I can do better than to describe the construction of some roads recently built in America in what may be called average conditions. I have already spoken of the roads which have been built under the recent road legislation in New Jersey. Those of Union County are particularly good, and I will briefly describe their construction.

Union County lies about 25 miles southwest of New York, contains about 100 square miles, and its population in 1890 was 72,467. Its main roads are 35 miles in length. Before their improvement was undertaken in 1890 most of them were mere mud tracks. On the passage of the legislation enabling the county to borrow money on its bonds for their improvement, a competent engineer, Mr. F. A. Dunham, was appointed to take charge of the work. He immediately made a survey of all the roads for the purpose of determining the gradients and the data necessary for preparing proper plans and specifications. When these were completed the work was advertised and let by contract to the lowest responsible bidder. The prices for grading were from 23 to 40 cents per cubic yard; for telford pavement from 80 cents to $1.12 per square yard. There was also a certain amount of work to be done in the way of drainage, culverts, bridges, etc. The width of the roads varied, according to the locality and the traffic, but the average width was 44 feet, with a crown or rise in the center of 12 inches. Of this width, 10 feet had a telford foundation, 14 feet had macadam metal, and two wings, 10 feet in width, on each side were of earth. The road was first graded to its approximate form, and then the space of 10 feet in the middle was excavated to a depth of 12 inches. This was then thoroughly rolled in order to compact the earth on which the stone portion of the road was to be built. Next the telford was laid. This consisted of irregular pieces of trap rock about 8 by 12 inches on the under side, 4 by 6 inches on the upper side, and 8 inches in height. These were placed by hand as closely together as possible, and the spaces in the surface were filled in with spalls and smaller pieces of stone, which were wedged into the openings as tightly as possible. A small amount of fine trap screenings was then spread over the telford for binding, and it was then thoroughly rolled. The macadam was placed over this in two layers, each of which was 2 inches thick, the first layer consisting of stone broken to 2 inches in size, and the second of stone broken to 1½ inches in size. Each layer was finished with a small amount of fine binding material, and then thoroughly rolled with a 10-ton roller, the surface being kept constantly wet by a sprinkling cart while the rolling was in progress. After the stone road in the middle was completed the earth roads on the side were rolled, and the road was finished.

These roads have given great satisfaction to all the residents in the county; they have been in use for several years with very slight repairs, and are still in excellent condition. They can be maintained so with proper care and at small cost for a long period. When the upper courses of macadam stone are worn out it will be necessary to resurface them with fresh layers of broken stone, and the road will then be in good condition for another term of years.

The cost of these roads was a little more than $8,700 per mile, and they may be taken as a type of the most expensive roads that it is necessary to construct anywhere outside of the boundaries of cities and towns.

In the southern part of New Jersey the roads have been constructed on a much smaller scale, the width of the road being about 20 feet and the metal portion only 8 feet; the telford foundation has been omitted, and the thickness of the macadam reduced to 8 inches. The cost of these roads has been about $3,000 per mile.

In New York some excellent roads have been built in the vicinity of Canandaigua, where the town bought a stone crusher and steam roller. They built "macadam roads consisting of a crushed-stone roadbed about 8 feet wide and nearly a foot deep in the center of a turnpike some 25 to 30 feet in width, sloping enough to shed the surface water, but not too deep to drive on any part of it, at an expense of $400 to $700 per mile, the smaller sum in cases where the stone had been contributed and deposited in piles by the neighboring farmers without expense to the town." The very low cost of these roads is due to the small cost of the stone; part of it was furnished free by the neighboring farmers, and all of it was obtained at a very low cost, not exceeding 20 to 32 cents per cubic yard.

While it is not possible to construct important highroads in the vicinities of large cities at any such price as this, yet it is possible to duplicate this work on ordinary country roads wherever the farmers are willing to cooperate. The plant necessary for the purpose consists of a portable stone crusher and steam engine costing about $4,000, and a steam roller costing about $2,500, or a total expenditure of $6,500. These should be owned by the county, and can be moved to any part of it where the road building or repairing is in progress. The broken stone can, in the great majority of cases, especially in the Middle and Eastern States, be obtained from the fields. It is an injury in the fields and a benefit on the roads, and all that is necessary is to collect it in the fields, haul it to the road, and deposit it in piles there. A large part of it is already broken to the proper size, at least for the lower course, and the rest of it can be run through the crusher at an expense of about 20 cents per cubic yard. The crusher can move along the road every night or twice per week, thus reducing the haul of the stone to and from the crusher to a minimum.

Long experience has shown that the only form of durable road is one made with crushed stones. There has been a difference of opinion among engineers as to the necessity for the telford foundation, but the generally accepted opinion now is that the telford should be used, and is worth more than it costs on roads of heavy traffic. On roads of light traffic, like those of Canandaigua, it can be omitted. The size of the stone can vary in the different courses if anything is to be gained by it, the larger stones, say up to 3 or 4 inches in size, being placed in the lower third, stone of about 2½ inches in the middle third, and stone of 1½ inches in the upper third. A small amount of stone dust or screening is necessary for a finishing coat. On the other hand, if any great expense would be incurred by separating the stone into different sizes this can be omitted, and the entire road made with stone not exceeding 2 inches in size. The minimum thickness should be 8 inches, and it should be laid in two layers, each of which should be thoroughly rolled by the heavy steam roller. Of the different kinds of stone the most durable is the trap rock or basalt, such as is found in the Palisades of the Hudson; the next most durable is granite and gneiss, then "flints" or quartz pebbles, then limestone, and finally sandstone. There is great difference in the hardness of different varieties of limestone, and the softer varieties, as well as the sandstones, are almost worthless for road purposes. Shaly and laminated stones should not be used, and stone containing a large amount of mica is also very undesirable. Oyster shells have been used in the construction of roads, but for the surface coat they are not suitable, as they quickly grind to powder. They can, however, be used to advantage in the lower courses. Gravel is also used, and it makes a road intermediate in durability between stone and earth. There is, however, a wide range in the quality of different gravels, some of them containing pure quartz pebbles and sand with a small amount of clay, all mixed in such proportions and of such size as to form a compact and very durable road when properly rolled down. There are some very remarkable beds of gravel of this character near Vicksburg and at other points in the Lower Mississippi Valley. Other gravels consist of soft-clay pebbles and a large proportion of clay, and roads made with these are very deficient in durability.

It is the practice on the Continent of Europe to carry the macadam from gutter to gutter, and in fact a great number of the highroads of France and Belgium are paved their entire width with heavy granite blocks. Such roads are durable, but otherwise undesirable. The practice is, however, universal in America, and it is a good one, to confine the stone or metaled portion of the road to a width varying from 8 to 16 feet in the center, and to have wings or earth roads on each side from 8 to 15 feet each in width. This arrangement not only reduces the first cost, but it furnishes earth roads which are traveled by nearly everyone in the summer on account of giving a softer material for the horses' feet. The wings are used not only for pleasure traffic, but in dry weather for hauling heavy loads. This use is beneficial to the earth wings, as it compacts and consolidates them, and thus affords a slope for the water to run off, and in addition it saves by so much the wear on the stone portion in the middle.

One of the most useful results accomplished by the Road Bureau in the Department of Agriculture is the collection of data in regard to road materials throughout the entire country, and the ability and willingness of the railroads to transport them at low rates. General Stone has carried on a correspondence with every important railroad in the country on this subject, and he has obtained answers from them which indicate not only a willingness but an active desire on the part of all the railroads to cooperate to the fullest extent in the improvement of the common roads. It is evidently to their interest to do so, because the common roads not only act as feeders to their own system, but in addition the bad roads cause them an enormous and unnecessary expense. Everyone who travels over a railroad must marvel at the enormous number of empty freight cars which he sees standing idle as he enters and leaves each town. These cars are counted by the hundreds of thousands, and every one of them represents from $300 to $500 of capital which lies idle during the greater part of the year. The principal reason for their lying idle is the bad condition of the roads. It is only during certain months that the farmers can haul their produce to the railway station, and then there is an enormous demand for cars, and all the cars, for a short period, are brought into use hauling a load one way and going back empty. After the rush is over the cars again lie idle for months.

Now, if the roads were in such condition that the farmers could deliver their produce regularly throughout the year probably one-third of the rolling stock could be dispensed with, and the trains in a majority of cases would haul loads both ways instead of coming back empty. The information obtained by General Stone makes it possible not only to determine at just what points in each State suitable road material can be obtained, but it shows that the railroads are willing to transport these materials at surprisingly low figures; some of them are willing to transport it free of cost, others at half usual rates, others at actual cost as nearly as it can be determined. As the result of all his inquiries, General Stone estimates that the average cost of moving broken stone by railroads would be about 2 mills per ton per mile; or, in other words, a cubic yard of broken stone weighing 2,800 pounds could be carried 100 miles for 28 cents, or for the cost of moving it about 1 mile on an average road by wagon. There is hardly any State, even in the prairies of the West, where a stone quarry can not be found within 200 miles by rail of any particular road to be improved, and the cost of 56 cents per cubic yard for transportation is by no means prohibitive. It would seem, therefore, as if broken stone suitable for road purposes could be brought within the financial resources of nearly every county in every State.

In some portions of the country there are tracts of sand and no stone is available. Fairly good roads can be made in such cases by mixing clay with the sand, if clay is available, and then rolling it. Some interesting cases are cited in Wisconsin where roads have been made by uniting shavings with the sand.

In the prairie districts of Illinois and elsewhere fairly good roads have been made in the following manner: The road is made by plowing two furrows 16 inches wide and about 12 inches deep under what are to be the wheel tracks, turning the earth inward, and two more for ditches also turned inward, which results in a slight raising of the roadbed, then filling the inner furrows with field stones or coarse gravel and finishing with a light coating of fine gravel.

Where the road runs through wet soils or springy places it must be thoroughly drained, or all work upon it will be thrown away. These drains can be made by excavating a trench about 15 inches wide and 1 to 2 feet deep in the center under the metaled portions of the road, and placing at the bottom of this a tile-drain or a rough stone box drain; the trough is then filled with stones from 2 to 5 inches in size. Where the road crosses a water course or other low point provision must be made for allowing these drains to discharge through the sides of the road.

Street pavements are only a special form of roads. They differ from ordinary roads in that they are designed to carry extra heavy traffic, and they are surrounded by houses containing a large population, so that questions of comfort, sanitation, and noise are considered, which can safely be disregarded in the case of roads through an open country. As in the case of common roads, so with street pavements, comparatively little attention was paid to the subject in this country until within the last fifteen or twenty years; and while the situation has partially changed and competent engineers have begun to study the problem, yet it is still true that the question of transportation within city limits has never received any such careful thought as has been devoted to every detail of the railroad problem. There is every reason to believe that if the streets were properly and smoothly paved a reduction in the cost of transportation within cities could be effected only inferior in magnitude to that which has been effected by the reduction of rates on the railroads of which I have already spoken.

The materials in common use at the present time for street pavements are stone blocks, asphalt, brick, wood, and macadam. Macadam is being rapidly excluded from use in city streets on account of its many objections. On any but the lightest traffic it is the most costly form of pavement that can be used in cities. In London and Paris macadam is being taken up and replaced by other materials as rapidly as possible on account of the expense of maintaining it, which was in excess of 60 cents per square yard per annum where the traffic is at all heavy. For a street 40 feet wide this is the equivalent of over $13,000 per annum per mile. Even where the traffic is not heavy the street is alternately mud or dust, according to the amount of water placed upon either in the form of rain or street sprinkling, and it is practically impossible to keep it properly cleansed. The wood pavement as laid in this country has everywhere proved a failure. It has sometimes been laid in the form of rectangular blocks, sometimes treated with a process for preventing decay and sometimes untreated. In other cases it has been used in the form of round blocks obtained by sawing off the trunks of the small, straight pine found in the States along the upper lakes. In this form it is very cheap, being frequently laid for less than $1 per square yard. But in all its forms it is lacking in durability, its life being about five years under ordinary traffic. In London and Paris the wood pavement has been laid in the form of blocks on a concrete foundation. The foundation is durable and the surface is pleasant to drive over, being smooth and noiseless when first laid, but it is not durable, and the cost of the foundation added to the cost of constant renewals of the wearing surface makes it very expensive. In general terms it may be said that any material of vegetable origin, subject to rapid decay, is unsuitable for a street pavement, both on the ground of lack of durability and on the ground of health.

The modern pavements, therefore, are of stone blocks, asphalt, or brick. Whichever is adopted should be laid on a concrete foundation, which is a permanent, durable structure, the wearing surface being a veneer which can be renewed from time to time as it is worn out by traffic. Stone-block pavements are generally adopted on steep grades and on streets which are subjected to heavy traffic. Trap rock, granite, and sandstone are the kinds of stone used for making paving blocks, but the latter are so deficient in durability that they are but little used, and the trap rock is extremely slippery. The principal stone pavements, therefore, are of granite, and the chief supply of them comes from the quarries of Maine, Massachusetts, and Rhode Island. The cost of granite-block pavements, with concrete foundation, is from $3 to $4.50 per square yard, depending upon the varying cost of labor and materials in different localities, and even in different parts of the same city. These pavements are durable and comparatively smooth during the first three or four years after they are laid; then if there is any traffic the edges are broken down and the surface becomes similar to that of cobblestones, and in this condition they will sustain traffic for twenty or thirty years. They are, however, rough and uncomfortable to drive over, and are very objectionable on the score of noise.

The asphalt pavements are of two kinds; one is a natural bituminous limestone found in France, Hanover, Sicily, and other parts of Europe, and consisting of about 90 per cent of limestone, in an impalpable form, and 10 per cent of bitumen. The material is crushed and ground to powder and then heated to a temperature of about 300°, taken to the street, spread and raked on a concrete foundation, and compressed by tamping or rolling. The thickness of the asphalt coating is from 2 to 2½ inches. This form of pavement was introduced into Paris about forty years ago, and it has since been laid in Berlin, London, and several other European cities. It has not reached a very great development because of its extreme slipperiness. It has been occasionally tried at different times during the last twenty-five years in America, but no large amounts of it have been laid on account of its slipperiness, and a very considerable portion of those that have been laid have been taken up and replaced by other materials. The other kind of asphalt is an artificial sandstone, consisting of about 90 per cent sharp silicious sand and 10 per cent of bitumen, which acts as a cement to bind the particles of sand together. The chief source of supply for this bitumen is a remarkable asphalt lake in the island of Trinidad, about 100 acres in extent and containing an apparently inexhaustible supply of the best quality of asphalt. This asphalt and the sand are separately heated to about 300°, a small amount of limestone is added, and the materials are then incorporated in a mechanical mixer, producing a uniform and homogeneous mixture of sand, limestone, and asphalt. This is then hauled to the street, spread, raked, and rolled in the same manner as previously described.

The gritty nature of this surface, owing to its sand constituent, renders it free from the objection on the score of slipperiness, and this form of pavement has obtained a wide development in America during the last twenty years, over 1,000 miles of it having been laid in upward of one hundred cities in the United States and Canada. It costs from $2.50 to $3.50 per square yard, according to the varying prices of labor and materials and the different thicknesses of foundation and surface, which can be varied to suit the traffic of any particular street. It is usually laid under a guarantee of five years, during which it is kept in order free of expense, and after that time it can be maintained for an apparently indefinite period in good order at an expense not exceeding 10 cents per yard per annum on streets of ordinary traffic. It is readily and easily repaired by simply heating the surface and adding fresh material, and under a proper system of maintenance, in which any defect is repaired the instant it appears, the surface is always in good order. The pavement is smooth, almost noiseless, durable, and if properly cleaned and occasionally washed can be kept cleaner than any other form of pavement.

The brick pavement has attained a rapid development in the last few years in different parts of America. It has been used in Holland for centuries, but its use on any considerable scale in America dates only from the last ten years. Its durability depends entirely upon the kind of clay which is used and the care with which it is burned. Some clays are incapable of producing a good paving brick, and in several Western cities the brick pavements laid with improper clays have gone to pieces in a few months. The burning should be stopped short of complete vitrifaction, otherwise the surface would be so glassy that horses could not stand upon it. On the other hand, if the burning is not carried far enough the bricks soon break up under traffic. The best clays for making paving bricks are found along the Ohio River in the vicinity of Wheeling, and in certain parts of Illinois. When made of proper clay and burned just to the right degree bricks make an excellent pavement, particularly for small cities. No special plant or facilities are required for laying them and they are very cheap in price. Including concrete foundation their cost varies from $1.50 to $2.50 per square yard, according to the locality and the distance which the bricks have to be transported.

And now one word in conclusion, and it should always be the last word in regard to roads of all kinds. That is, the necessity for prompt and systematic repairs. Therein is the secret of the success of the French roads, and it is the observance of this principle on railroads which makes possible high speed, comfort, and low rates. In America this principle has been wholly disregarded on the common roads and but slightly observed on the city streets. Any road, railroad, or pavement is subject to incessant pounding and rubbing. This is what it is built for. Now, this pounding and rubbing must inevitably produce wear, not only on the road itself, but on the vehicles and animals which pass over it. That road is the best which produces the smallest amount of aggregate wear or damage to the road and vehicle combined, and the greatest source of economy both for the road and the vehicle is the prompt repair of any defect the moment it appears.

ROAD BUILDING IN THE UNITED STATES.

By ROY STONE,

cial agent in charge of Office of Road Inquiry, United States Department of Agriculture.

[Address before the students of the Massachusetts Institute of Technology.]

GENTLEMEN OF THE MASSACHUSETTS INSTITUTE OF TECHNOLOGY: We have long heard of abandoned farms, and we are beginning to hear of factories abandoned, through the process of natural selection and the "survival of the fittest;" but in one of my recent journeys to the West I discovered a more startling subject—an abandoned railroad, with rusting rails, rotting bridges, a weed-grown roadbed, and even its right of way lapsing through nonuse.

This road led from a large and prosperous town through a rich farming district; it connected with other railroads at both ends and had no rival line, yet it had not business enough to pay operating expenses, even as a feeder to its connections, and I learned that this was one of several roads in the same condition.

Here is an object lesson too plain to be misunderstood. We have overdone railroad building in this country as a means of development, and must search for other methods of bringing the country up, even to the point of making its present railways profitable. In its bearing on the question of employment for engineers that lesson has doubtless come home to many instructors in engineering, and even their pupils must be casting anxious eyes into the future to see what employment is to replace railway building and give suitable occupation to the ceaseless stream of graduates in the art of construction. And were it not that, to those who are properly equipped, human progress opens constantly new fields for human effort, there might be room for serious apprehension in the sudden extinction of so large a share of the world's demand for constructive skill.

Among the growing fields for the employment of civil engineers may be reckoned ship canals, water powers, irrigation, and highways; and of all these, the last gives much the most hope of filling the place of railroads in this regard.

Many institutions are therefore awaking to the need of instruction in this branch of engineering. There is need, however, in this country, and in this particular field, for something more than mere technical instruction. Under a paternal government the engineer needs only to know how to build roads; the authorities do the rest. Under a popular government he must know how to get them built. His duty as a citizen and his professional interest alike require that he should know all about the laws and ways and means, and all the facts and arguments bearing on the question, and be as well able to promote road improvements as to execute them.

Professor Swain is therefore wise in proposing that one of my lectures shall "cover an account of what has been done in the United States in the past few years in improving the condition of the roads."

But before entering upon the question of what has been done by the various States I will speak of what has been done by the General Government. It is not commonly known that the Government was engaged in building highways for seventeen years before it began the improvement of rivers and harbors, and that the construction of national roads has been kept up to a greater or less extent ever since.

Everyone has heard of the magnificent Cumberland road, which crossed the Alleghany Mountains and led to the Mississippi, but few are aware that while this road was in progress twelve national roads were laid out in the States and Territories, making what was regarded then as a complete system of highways, and that more or less work was done in opening and constructing these various highways.

The monetary crisis of 1836 put an end to all great schemes of Government expenditure, and from that time to this only a few military roads have been made, and of late years only those leading to national cemeteries.

The constitutional question as to the power of the Government to build roads has been frequently raised but never fully decided, and Congress has continued making appropriations for the purpose for ninety years. Many people believe, therefore, that there would be little in the way of a more extended use of national aid to road building, so far, at least, as regards military and post roads. The necessity for military roads has been insisted upon by many army officers, in view of the difficulty of moving troops by rail for the suppression of domestic disorder at a time when railway employees are in sympathy with the malcontents, a contingency which in late years has frequently happened.

In addition to the direct road building done by the United States under contract, grants of land have from time to time been made to States in aid of road building, and the labor of United States troops has been employed in the work.

Returning to road building by the States, there is wide variation in their systems of legislation for this purpose, but a great amount of road building is going on in a miscellaneous way and much valuable knowledge is being developed.

Naturally, the first method considered in most States is the building of State roads, somewhat as Massachusetts has begun. This plan has met with little favor, however, in other States, and it is feared that local jealousies will prevent any good practical results from this method.

It is impossible for the State to build any large proportion of the roads that will be deemed necessary, and the people who are disappointed in getting the accommodation hoped for will soon be in the majority, and the State appropriations be liable to be cut off.

A plan, which came very near being adopted in New York State, of connecting all the county seats by a network of roads is apparently more fair than one which leaves a State commission to make a selection of those lines which seem to it most important. But this again is open to the objection that county seats are frequently not centers of business or travel, and long stretches of useless road would be built, while many business centers which are not county seats would fail to be accommodated. The same money spent in completing the same length of road in each county, but so located as to radiate from shipping points and other business centers, and to connect towns and villages not joined by railroad, would be of much greater benefit; if the State contributions were generally used, as they are beginning to be in New Jersey, to stimulate local effort and induce the expenditure of twice or thrice as much by counties, towns, and neighborhoods, the benefit would be much greater still, while all ground for local jealousies would be removed by the rule of "first come, first served," and of helping those first who are first ready to help themselves. The offer of State aid moreover induces more or less of a scramble for its benefits and does more to agitate the public mind on this subject than any other method of procedure. In New Jersey, farmers who were opposed to any movement for road improvement a few years ago are among the most anxious to have it brought to their own doors.

The State-aid roads, being widely scattered, form numerous centers of education, and soon the whole State will be ready for more rapid measures of road improvement. This effect is not bounded by State lines. People are coming from other States to see what is being done in New Jersey, and return to spread the news wherever they go that farmers are prospering even in the midst of these hard times, solely by reason of having good roads in place of bad ones.

A committee of the New York State legislature, together with members of the boards of supervisors, numbering nearly one hundred, visited some of the State-aid roads in New Jersey last spring and returned to New York and passed a bill for State aid through the lower house of the legislature by a majority of 4 to 1.

Another committee from Asheville, N. C., visited New Jersey, and on their return home their county entered at once upon a general scheme of road improvement.

These visible effects of a practical demonstration have led me to search constantly for any available means of starting road improvement, regardless of its being the best means possible. It is almost useless to carry on a campaign of education by mere talk and literature without something in the way of practical illustration and proof; hence I have endeavored to encourage every plan that leads to practical and present road building. The times are most unfavorable to any method which involves an increase of direct taxation, and in most rural localities there is an insuperable opposition to the issuance of bonds, notwithstanding that all those counties and townships which have heretofore borrowed money for road building have found their values increase so rapidly by the influx of outside capital that their rate of taxation has been diminished instead of increased.

There are, however, a few places where the direct tax method is being employed, and a considerable number where counties are issuing bonds, notably, in the first case, the town of Canandaigua, in the State of New York; and, in the second case, the county of Morris, in the State of New Jersey. Morris County has lately borrowed $350,000 at less than 4 per cent interest, and the townships of the county will tax themselves for half as much more for a complete system of county roads. In Canandaigua the property owners have paid an extra tax of $4,000 annually for several years, and are so well pleased with the result that they are likely to continue the tax until all their roads are macadamized.

The direct-tax method must necessarily be slow in its results, and for a long time it will be unequal in its benefits, since large portions of a township must pay the additional tax for years before their own roads can be reached; moreover, they will meantime be actually injured by the advantage given to their competitors in reaching their common markets.

The direct-tax method tends to secure economy in construction, but, by limiting the means, it narrows the scope of the improvement, and in some cases almost fatally. In Canandaigua, for instance, the authorities were obliged to choose between macadamizing the roads and lowering their grades, which latter required relocating and cutting down the hills; not feeling able to do both they chose to macadamize the roads as they were, and while they have made a great improvement they have in some cases probably barred out a greater one forever; when roads have been resurfaced at what appears to be great expense it is extremely difficult to make any subsequent change in their location. Their ideas of economy, moreover, forbade the employment of an engineer, and though they have made excellent roads they would have done much better under scientific advice. The fact, however, that they have succeeded with only the ordinary educated farmer's knowledge of road building shows that engineers, to make themselves useful and requisite in the construction of ordinary country roads, must be prepared to adapt themselves to the local conditions, and more especially must know all the possible economies of road construction.

There is a constant apprehension in the rural mind that the employment of an engineer or an architect leads to wasteful expenditure, and unless engineers fit themselves to make the best of the conditions existing they are liable to be left out in the great majority of cases of country road building, in which case both roads and engineers will suffer.

This is the particular branch of road building science and practice which I desire to develop, my special function in the Department of Agriculture relating to the building of country roads as distinct from city or suburban streets, and I am extremely anxious that engineers who are intending to devote themselves to road construction should be able to give the benefit of scientific knowledge to the construction of roads even of the simplest and cheapest character.

Returning to road legislation and the definite question of ways and means, the county bonding laws passed by various States have been useless where they require a popular vote to authorize the issuance of bonds; they have been effective only in a few cases in New Jersey and in some counties in other States where a vote was not required. It is very clear, then, that in no part of the United States are the masses of the people educated up to the point of being willing to tax themselves or to borrow money to any great extent for the purpose of road building. In Missouri, where an active campaign for road improvement has been carried on for several years, the people have lately voted down, by a large majority, an amendment to the constitution merely permitting counties to vote a special tax for this purpose; yet, while in no State, nor scarcely any county, is there a sufficiently developed sentiment in favor of road building to secure a majority of votes for any extensive measure of practical improvement, in almost every county in every State there is some neighborhood far enough advanced to be ready to contribute a share toward the improvement of its roads if the county and State will do the rest.

Apparently, then, this opens the only practicable way for immediate action in all the States. Accordingly, legislation in favor of State aid is becoming a favored scheme among the advocates of good roads; and it is through State aid only that the cities and villages, which are generally quite ready to assist in road building, can make that assistance available.

In Ohio, Indiana, and some other Western States improved roads are built by assessing a part of the cost on a strip of 1, 2, or 3 miles in width on either side of the road, and the balance upon the county at large, but this plan is not wholly just; in fact, the benefits of road improvement will rarely if ever be bounded by parallel lines.

Roads that are important enough to require systematic reconstruction will generally be such as radiate from railway stations or boat landings, market towns, county seats, or villages, and the district benefited by each will widen rapidly as the distance from the common center increases. Taking the case of four main roads diverging from a railway station at right angles to each other, these roads at a distance of 5 miles will be 7 miles apart, and the benefit district of each will be a triangle with its apex at the station and a width of 7 miles at the base. The bounds of the district, moreover, will be modified by many such obstacles as streams, swamps, hills, valleys, etc., which will divert travel to or from a particular road, or by artificial conditions, like the location of factories, creameries, grain elevators, schools, or churches. All these conditions, however, are susceptible of a fairly exact determination, and the benefit district of a road can be almost as accurately defined as the drainage area of a stream. When it is so defined such a district forms an ideal unit of action for road improvement; the interests of all its people are identical, though unequal, and the share of expense each ought to bear can be safely left to the commissioners charged with its assessment.

The bill passed by the assembly of the State of New York divided the total cost of roads built with State aid into three equal portions, one to be paid by the State, one by the county, and one by the benefit district, except in cases where the benefit district was peculiarly restricted and its area fell below an average of 1 mile wide on each side of the road, in which case its share was to be scaled down proportionately to its decrease in area, but not in any case below 10 per cent of the whole. The local assessment upon the benefit district was made payable in ten annual installments, with interest, the county meanwhile advancing the money. In the case of roads built from local materials, as in Canandaigua, and made with single track of stone and a parallel track of earth, which is now considered the best practice for farming districts, the total cost will approximate $1 per acre on the 2 square miles of area assumed to be benefited on the average by each mile of road, of which amount only 33 cents per acre falls upon this area, which, divided over ten years, amounts with interest to only 4 cents per acre annually.

The present customary road tax averages about 10 cents per acre annually, and if the increase of 4 cents per acre is found too heavy, the present tax might safely be reduced to 6 cents per acre for road repairs, leaving the total cost no greater than at present. The roads in Canandaigua have cost about $900 per mile, but it would have been better economy to have added a few hundred dollars to this cost.

I have said enough on the subject, perhaps, to show that, through the application of State aid, good, permanent roads can be built without increasing the burdens of the farmer, and all experience has shown that the expense to the State and county is much more than balanced by the increase in taxable values of the districts through which the roads are built.

Another method of road building which is coming much into vogue, especially in the Southern States, is through the use of convict labor, and so much is being accomplished in this direction that it is worth while to give careful consideration to its details.

There are two sides to the question of working convicts on the highways, or rather two sides and a broad middle ground. The negative side is taken by the prison association of New York and by penologists generally. The reasons advanced in opposition to the plan are that honest labor would be interfered with; that a large body of keepers would be required at great expense; that there would be a constant necessity for shooting convicts in order to prevent escapes; that in many cases the prejudice against convict labor would require a military force to protect convicts thus employed; and it would be demoralizing to the convicts themselves to employ them in public places.

This is a view of the question natural to men whose minds are fixed on the need to society of the reformation of criminals. Opposed to it is the opinion of many good citizens who seek the public advancement in other ways, and especially in the direction of improved means of communication, and who see in the convicts, now idle in our jails and prisons, a labor force sufficient to mend all the roads in the country if it could be so applied, and which they believe could be so applied without prejudice to free labor. The advocates of the convict road work insist further that the outdoor life and exercise afforded by such employment would benefit the health and morals of the prisoners.

In the vicinity of Charlotte, N. C., convicts have built miles of substantial roads, and with such satisfaction to the people that the special law under which it was done is now being extended to other counties. In other Southern States, where the convict-lease system still prevails, it is clear that a transfer of the prisoners from irresponsible and often inhumane private employment to the care of States or counties would be a saving kindness to them and would benefit the entire community.

Some of the apprehensions of the New York Prison Association do not appear to have been well founded. The legislature passed a bill providing for the employment of convict labor on some of the roads of the State, in spite of the protest of the association, and a very satisfactory experiment was made at Clinton Prison. There was no interference with the convicts by the citizens except in two cases, where intoxicated men offered them liquor; no apparent demoralizing effect on the prisoners or the public; no shooting of convicts, and only three men attempted to escape. In his report on the subject the warden of the prison concludes as follows: "That a limited number of convicts can be worked successfully is now an established fact."

On the other hand, however, when we examine the warden's financial statement, we find but little, if any, economy in the use of convicts, as compared with the employment of free laborers for the same work. The cost of maintenance and guards and of the search for escaped convicts was equal to 91 cents for each day's labor done, which, considering the comparative efficiency of such labor, is very near its full value, a day's work being only eight hours.

Again, it may be safely predicted that when road making becomes a great business in the country the introduction of labor-saving appliances will do away with a large share of the hand labor now required in laying a stone or gravel road. The material, being generally transported by railroads, will then be transferred to wagons without shoveling, and from the wagons will be mechanically spread in its place, so that almost nothing will be left for convicts to do on the line of the road.

These considerations strengthen the position of those who hold the middle ground of the question, which is that State prisoners should be employed wholly in the preparation of road material, and in places where they can be as easily guarded and secluded as in the prison. The plan proposed for this is in substance as follows: First, for the State to buy some of the territory which contains the best rock within its limits; second, to make the necessary railroad connections, having first secured the permanent agreement of all its leading railroad companies to carry road material at the cost of hauling, on condition, if required, of the State furnishing to them a certain amount of track ballast free of charge or at the cost price; third, having erected the necessary buildings and provided the best machinery for quarrying and crushing rock, to bring all able-bodied State convicts and put them at this work; fourth, the counties to put their jail prisoners and tramps at the work of grading, draining, and preparing the road for macadamizing; fifth, the State to furnish broken stone free on board cars as its contribution to road improvement.

The cost to the State, in addition to the maintenance and guarding of the convicts, would be only that of fuel and oil, explosives, and use of machinery, or, according to the Massachusetts commission's report, 6.8 cents per cubic yard of broken stone, amounting, for the 1,200 cubic yards required to lay a mile of single-track road 9 feet wide and 8 inches deep, to $81.60.

The remaining cost would be the railroad freight, amounting, for an average distance of 100 miles, to not more than 28 cents per yard, or $336 per mile; the wagon haul, averaging possibly 2¼ miles, 30 cents per yard, or $360 per mile, and the rolling, superintendence, and incidentals (not including engineering, which woulbe a general county charge) 10 cents per cubic yard, making the total local cost 68 cents per cubic yard, or $815 per mile.

This plan would bring the expense of road improvement so low that no elaborate scheme of taxation or borrowing would be necessary, and all its benefits could be speedily and universally realized. The best plan for carrying it out would perhaps be to let the "benefit district," as heretofore defined, pay one-third of the cost by installments, and the township one-third; the county to pay the remainder, and to advance the amount for the district, with a rebate or discount to all individuals who might prefer to pay in cash, so that no one would be put in debt against his will. The cost to the benefit district on this basis of division would be $272 per mile.

The growth of the road movement in North Carolina is unquestionably due to the use of convicts. Indeed, this use has in the majority of cases been the most important factor in deciding the counties of that State to vote a tax for the improvement of the public roads. The result of the experiments in North Carolina has been altogether favorable to the system, both in point of efficiency and economy and in respect of the health of convicts. In Lenoir County only short-term convicts are employed. They are carefully described and photographed, and offered certain inducements in the way of reward or shortening of term if they remain at their post and faithfully discharge their duties. With these precautions they are employed on the public roads very much as hired laborers would be, under the control of a superintendent, but without a guard, and they are allowed to remain at their homes from Saturday night until Monday morning. This novel experiment has been in operation a year, and not a convict has attempted to escape.

Many of the States are now arranging to establish supply camps for road material at which State convicts can be worked under proper restrictions. California, especially, will profit by this change; the State prison grounds at Folsom contain an excellent vein of trap rock and abundant water power, with the necessary machinery for moving and crushing rock; there are 700 or 800 idle convicts in the prison, who could all be employed in this manner, and the prison authorities find that they can prepare road material and put it on the cars for 20 cents a cubic yard, and pay all of the expenses of the prison out of the receipts.

Many railroads have already offered to transport material for the bare cost of train service, and probably most of the railroads in the country would be willing to make such an arrangement, for they are naturally interested in the improvement of the highways. Their deep concern in the condition of its highways is universally recognized; it was particularly well expressed at the Michigan engineers' convention by Mr. E. W. Muenscher, chief engineer of the M. and G. R. Railway, who said: "No interest in the State of Michigan would be more benefited by good roads than the railroads. During a large part of the year much of their rolling stock is lying idle because farmers can not bring their produce to the station. At other times they can not get cars enough to haul away this produce, and the sidings, elevators, and warehouses are gorged to overflowing; extra forces must be employed; men in other lines of traffic who need cars are denied, and general disturbance of business, delay, and loss follow. Good roads would distribute this business more uniformly over the year, to the mutual advantage of the companies and their patrons, to say nothing of the increase of population and production and prosperity which would follow."

In the State of Kentucky very excellent roads have been built under special laws permitting counties to contribute $1,000 per mile to the construction of stone roads of specified details, and the State has now a larger mileage of good roads in proportion to the population than any other in the United States.

The various methods in practice consist, therefore, of: First, direct State road building, as in Massachusetts and some of the Western mountain States; second, State aid, as in New Jersey; third, county road building; fourth, county aid to local contribution; fifth, convict labor; sixth, direct taxation; seventh, assessment of benefit strips; eighth, township bonding, as in Pennsylvania. Under these various methods roads enough will undoubtedly be built so much better than those at present existing in this country as to whet the appetite and pave the way for building better and more costly roads in the future; and while I urge on young engineers to be prepared to build these cheaper roads chiefly for their educational advantages to the general public, I do not by any means desire to lower the ultimate standard of road building. The great highways of the future will assuredly be better structures than anything in the present or the past, and fully worth all a trained engineer's most earnest study and attention. Undoubtedly the wearing surfaces of stone or asphalt will be replaced by flat steel rails suited to all vehicles and laid level with the roadway so that wheels will pass onto or off them without difficulty, a slight rise at the edges giving only a gentle guidance to the wheels.

There is no more comparative economy in running a wagon over a rough surface than there would be in running a railway train on the ground, and the saving in wear on roads of heavy travel, with the present low price of steel, would alone more than justify the additional cost of construction, while the saving in traction would be from 50 to 80 per cent.

These great highways will connect all the large cities and will be crowded with vehicles of kinds now seen only in exhibitions or on trial trips. One need not look far into the future to see horses superseded for all road uses except pleasure, and largely for that. In the race for automobile vehicles between Paris and Rouen last July over a hundred vehicles of different descriptions were entered. Those driven by electricity were barred out by the want of facilities for recharging. In the next race these vehicles will be provided for, and with the steady diminution in weight required for storage batteries they will undoubtedly be a favorite class.

The recent extended development of cheaper water powers, especially in this country, together with the success achieved in the transmission of power by electricity, promises an abundant supply of motive power for such vehicles on all the great highways of the future at a cost far below that of animal power.

The city of Augusta, Ga., furnishes water power at the rate of $5.50 per horse power per annum. This is at the rate of 1¼ cents per day. Supposing this to be doubled for transmutation, and doubled again for transmission and waste, and still again for profit, it would still furnish power along the public roads for $1 per week per horse power.

One of the latest of the electrical carriages for two persons weighs only a little over a ton, including passengers and battery. One horse power will move this vehicle over a good stone road at 15 miles per hour, or 1,000 miles in a week by daylight. This gives a journey of 1,000 miles for two persons for $1, or at the rate of about one-fortieth of the cost of railway travel.

This is progress enough in the art of transportation to suffice for a few years at least. The students of aerostation are succeeding so well in the art of flying that when they do equally well in the art of alighting they may do away with the need of roads altogether, but until then we must go on improving both roads and vehicles and making traveling as near like flying as possible.

●

www.ingramcontent.com/pod-product-compliance
Lightning Source LLC
Chambersburg PA
CBHW022017190326
41519CB00010B/1547

* 9 7 8 3 7 4 4 6 4 8 7 5 2 *